The Mathematics Enthusiast

Volume 13
Number 3 *Number 3*

Article 1

8-2016

TME Volume 13, Number 3

***TME* Vol.13, no.3 (August 2016) pp.187-314**

"Zachor!"(Remember!)- Elie Wiesel (1928-2016)

Editorial

Bharath Sriraman[1]

University of Montana – Missoula

This issue of The *Mathematics Enthusiast* has a perfect number of journal articles! The six articles range from topics in mathematics [Number theory, Complexity of Algorithms, Calculus] to topics in mathematics education [problem solving and aesthetics] and one on mathematics philosophy which examines the foundations of modern theorem proving.

The secret life of 1/n forges deep connections between number theory and algebra as well as analysis. It is the feature article of this issue for the simple reason that even a seemingly simple experimental observation in mathematics can be pursued to substantial depths- in this case decimal expansions connections to primitive roots and class numbers, something any mathematics enthusiast would appreciate. A reader that has encountered André Weil's slim volume "*Number Theory for Beginners*" might appreciate the contrast of Lyon's approach to that in the aforementioned book, where primitive roots are introduced using a strictly algebraic approach on p.47. Interestingly enough Weil's book was based on the compilation of notes by Maxwell Rosenlicht (his assistant at that time) from a 10-week introductory course in number theory taught at the University of Chicago in 1949. Suffice it to say that anyone who works through Weil's book today is ready for a graduate course in algebraic number theory. Similarly the reader who digests Lyon's approach would appreciate the necessity of learning algebra!

The five other articles in this issue are also delightful reading. We hope you enjoy this issue.

Reference

Weil, A. (1979). *Number theory for beginners (with the collaboration of Maxwell Rosenlicht)*. Springer-Verlag, Berlin.

[1] sriramanb@mso.umt.edu

The Mathematics Enthusiast, **ISSN 1551-3440, vol. 13, no.3**, pp. 187 – 188

The secret life of $1/n$: A journey far beyond the decimal point

Christopher Lyons
California State University, Fullerton

ABSTRACT: The decimal expansions of the numbers $1/n$ (such as $1/3 = 0.3333...$, $1/7 = 0.142857...$) are most often viewed as tools for approximating quantities to a desired degree of accuracy. The aim of this exposition is to show how these modest expressions in fact deserve have much more to offer, particularly in the case when the expansions are infinitely long. First we discuss how simply asking about the *period* (that is, the length of the repeating sequence of digits) of the decimal expansion of $1/n$ naturally leads to more sophisticated ideas from elementary number theory, as well as to unsolved mathematical problems. Then we describe a surprising theorem of K. Girstmair showing that the digits of the decimal expansion of $1/p$, for certain primes p, secretly contain deep facts that have long delighted algebraic number theorists.

Keywords: Decimal expansions, rational numbers, primitive roots, class numbers

The Mathematics Enthusiast, **ISSN 1551-3440, vol. 13, no. 3**, pp. 189-216

Introduction

If we spend enough time punching numbers into a calculator—say, as students in a science or engineering class with a lot of numbers to crunch on our weekly homework assignments—it's likely we're going to start recognizing certain decimal expansions, whether we choose to or not. For instance, I'd wager that most readers are familiar with the following decimal expansions:

$$\begin{aligned} 1/2 &= 0.5 \\ 1/3 &= 0.3333333... \\ 1/4 &= 0.25 \\ 1/5 &= 0.2 \end{aligned}$$

Many will also recognize

$$1/6 = 0.1666666...$$

But how about the expansion of 1/7? I think we've all seen it:

$$1/7 = 0.142857142857142857...$$

But perhaps the "repeating part," which is the 6 digit sequence 142857, is a little too long to stick in most people's memories unless they intentionally put it there. Similarly, some of us know the expansion:

$$1/11 = 0.0909090909...$$

but if we boost the denominator up by 2 then most probably haven't memorized

$$1/13 = 0.076923076923076923...$$

The most common view of decimal expansions like the ones above is completely pragmatic: they help us approximate certain quantities to a desired degree, and perhaps some of the digits (if we're in a situation that also involves a bit of random error, such as a scientific measurement) are not to be entirely trusted. In these settings, one might look at the examples above and perhaps use something between 2 and 10 digits out of the whole decimal expansion, and then discard the rest through the usual process of rounding. But what if we were to pay attention to *all* of the digits past the decimal point? Is there anything more interesting that these decimal expansions have to offer besides their role as everyday computational workhorses?

In this exposition, we'll explore this question for the decimal expansions of the very simple class of numbers represented above, namely $1/n$ for integers $n \geqslant 2$. These might seem like incredibly mundane strings of digits, especially if one compares them to the decimal expansion of π, which is the object of much computational and popular attention. Yet as we'll see, even the simplest questions about the expansion of $1/n$ lead to unexpectedly sophisticated ideas, and to problems that have stumped mathematicians for at least two centuries and remain unsolved even today. We'll also discuss a discovery of Kurt Girstmair from the 1990s—which of course is relatively recent in the long history of our use of decimals—that shows how some of these expansions secretly contain information about certain integers that have been studied in number theory since at least the early 1800s. So in fact the decimal expansion of $1/n$ is often far from mundane!

Here's a brief overview of the contents of this article.

In Part 1, we focus on the overall structure of the decimal expansion of $1/n$, without paying much heed to the actual digits that appear. In §1.1, we determine those values of n for which the expansion of $1/n$ consists purely of a sequence of ℓ digits that repeats forever; in this case we call ℓ the *period* of the decimal expansion of $1/n$. For a given n, it may not be clear how large ℓ can be, and in §1.2 we place a basic restriction on its size, meeting a strange-looking function along the way called *Euler's ϕ-function.* We digress for a bit in §1.3

to discuss those values of n for which ℓ is as large as possible; this brings us face-to-face with the curious idea of a *primitive root* in number theory, and with an enduring mystery. We return in §1.4 to the general question of how we can determine ℓ in terms of n, by making a significant improvement to our result from §1.2. In §1.5, we finally see what lies at the very core of this question.

Part 2 is shorter than the first, and also less detailed due to the depth of some of the topics it surveys. In this part, we return to the expansions studied in §1.3 for which the period of $1/n$ is as long as possible, and we look more closely at the repeating strings of digits they involve. In §2.1, we discuss Girstmair's result, which shows that in spite the random-looking appearance of these strings of digits, they can actually encode the values of special number theoretic quantities that were first introduced by Gauss. We describe these quantities in §2.2, after surveying a small portion of the theory of *binary quadratic forms*, a venerated and seemingly far-flung domain of 18th and 19th century number theory.

In Part 3, we quickly mention some more general statements and situations that have been omitted in the previous parts, and give some suggestions for further reading.

The Appendix contains a proof a result quoted in §1.5

To finish this introduction, let me say that this article has its origins in a talk that I've given to several undergraduate audiences. For this reason, I've chosen to keep the tone somewhat informal, to keep the pace relaxed, and to give plenty of examples. Nothing in this exposition is new, especially since the subject of periodic decimal expansions has piqued the interest of many for centuries (see [Dic] for a detailed account). There are many articles on various aspects of periodic decimal expansions that have been written for a fairly general mathematical audience, such as [JP, Lea, Ros, SF, Gir3] which are of varying levels and have some intersection with the topics discussed here. Yet most of the natural questions explored here are still unknown to even the average mathematics undergraduate, despite being understandable to anyone who has a basic familiarity with the idea of an infinite decimal expansion. In a world where mathematicians are often asked why math doesn't stop after calculus, one of the underlying motivations here is show how *nontrivial questions, sophisticated ideas, and even unsolved problems can arise from mathematical objects that are as commonplace as the decimal expansion of* $1/n$.

It's my pleasure to thank Bharath Sriraman for inviting me to turn an earlier set of notes on this topic into the present article. I also thank the various undergraduates who have listened to me speak on some of these topics and helped improve my exposition.

1 How long is the decimal expansion of $1/n$?

This part will explore the question in the title, in more than one sense. Before getting into things, I want to set down two items of notation. The first is a handy device that you've probably seen before: if a sequence of digits in a decimal expansion repeats forever, we'll indicate this by putting a line over that sequence. Three examples are:

$$1/3 = 0.\overline{3}, \qquad 1/6 = 0.1\overline{6}, \qquad 1/7 = 0.\overline{142857}.$$

The second notational device will help in situations where a decimal representations might be confused with a product. Usually we see the expression $d_1d_2d_3$ and interpret it as the product of the quantities d_1, d_2, and d_3. For this reason, I'll often use a `typewriter font` in instances where I'm viewing the variable as a decimal digit. In particular, when I write $\mathtt{d}_1\mathtt{d}_2\mathtt{d}_3$, this should be viewed as the usual decimal representation of the integer $\mathtt{d}_1 \cdot 10^2 + \mathtt{d}_2 \cdot 10 + \mathtt{d}_3$; thus if $\mathtt{d}_1 = 3, \mathtt{d}_2 = 7, \mathtt{d}_3 = 5$, then the expression $\mathtt{d}_1\mathtt{d}_2\mathtt{d}_3$ represents the integer 375 and *not* $3 \cdot 7 \cdot 5 = 105$.

1.1 Some preliminary steps

At some point in our mathematical education we come across the fact that, when we look past the decimal point, the decimal expansion of a rational number is either finite in length or it eventually repeats the same finite sequence forever. If we look specifically at the rational number $1/n$, when is its decimal expansion finite and when is it infinite? If it's infinite, when does it start out with some initial "non-repeating part" before reaching a repeating sequence (such as $1/28 = 0.03\overline{571428}$) and when is it "purely repeating" (such as $1/41 = 0.\overline{02439}$)?[1] The answer is summarized in the following:

Proposition 1.1. *Let $n \geqslant 2$ be an integer.*

(a) The decimal expansion of $1/n$ is finite if and only if n is not divisible any primes other than 2 and 5.

(b) Suppose the decimal expansion of $1/n$ is infinite (meaning n is divisible by at least one prime other than 2 or 5). Then the decimal expansion is of the form

$$1/n = 0.\overline{\mathtt{d}_1\mathtt{d}_2\ldots\mathtt{d}_\ell}$$

for some digits $\mathtt{d}_1, \ldots, \mathtt{d}_\ell$ if and only if n is divisible by neither 2 nor 5.

So looking at part (b), we see that the non-repeating portion of the decimal expansion of $1/28$ is explained by the fact that $4 = 2^2$ divides 28, while $1/41$ is purely repeating because 2 and 5 don't divide 41. I'll also note that the condition in part (b) can be stated with fewer words: instead of saying that n is divisible by neither 2 nor 5, we can say that the greatest common divisor of n and 10 is 1, which we write as $\gcd(n, 10) = 1$. We'll often use this more concise phrasing below.

Now why are the statements in Proposition 1.1 true? I don't want to dive into a full proof here, but some examples will help highlight the most interesting points. To demonstrate one direction of the statement (a), look at $n = 4000 = 2^5 \cdot 5^3$: we have

$$\frac{1}{4000} = \frac{1}{2^5 \cdot 5^3} = \frac{5^2}{2^5 \cdot 5^5} = \frac{25}{10^5} = 0.00025.$$

This trick will work in general: if $n = 2^a \cdot 5^b$, then we can write $1/n$ as some integer divided by a power of 10 (which is what it really means to have a finite decimal expansion!), simply be "adjusting" the number of 2's or 5's in the denominator to get a power of 10 and then compensating in the numerator. Looking at the other direction of (a), if the expansion of $1/n$ is finite, then this means $1/n = m/10^k$, and hence $nm = 10^k$. So a prime dividing n also divides 10^k, implying that prime is either 2 or 5.

The statement (b) is more interesting to think about because it gets to the heart of what an infinite decimal expansion is: a convergent infinite geometric series. Consider the infinite decimal expansion $0.079\overline{54}$. Writing it more verbosely as

$$\begin{aligned} 0.079\overline{54} &= 0.079 + 0.00054 + 0.0000054 + 0.000000054 + \cdots \\ &= \frac{79}{10^3} + \left(\frac{54}{10^5} + \frac{54}{10^7} + \frac{54}{10^9} + \cdots\right), \end{aligned}$$

we see that the quantity in parentheses is an infinite geometric series whose first term is $54/10^5$ and whose ratio is $1/10^2$. So by the main theorem on infinite geometric series, we can conclude

$$0.079\overline{54} = \frac{79}{10^3} + \frac{54/10^5}{1 - 1/10^2} = \frac{7}{88}.$$

This number is of course not of the form $1/n$, but it is rational and makes the point that infinite decimal expansions that are eventually periodic are always equal to rational numbers.

[1] At this point, I'm hoping these notions of "finite" vs. "infinite" decimal expansions, and "non-repeating parts", are intuitive enough. But see §1.2 for a more precise definition of these terms using the *shift operator*.

Moreover, after studying this example carefully, one can also see how similar steps will show that

$$0.\overline{\mathtt{d}_1\mathtt{d}_2\mathtt{d}_3\ldots\mathtt{d}_\ell} = \frac{m/10^\ell}{1-1/10^\ell} = \frac{m}{10^\ell - 1}, \tag{1.1}$$

where

$$m = \mathtt{d}_1 \cdot 10^{\ell-1} + \mathtt{d}_2 \cdot 10^{\ell-2} + \cdots + \mathtt{d}_{\ell-1} \cdot 10 + \mathtt{d}_\ell,$$

i.e., m is the positive integer that in decimal notation is written as $\mathtt{d}_1\mathtt{d}_2\mathtt{d}_3\ldots\mathtt{d}_\ell$. So if $1/n = 0.\overline{\mathtt{d}_1\mathtt{d}_2\mathtt{d}_3\ldots\mathtt{d}_\ell}$, this implies $nm = 10^\ell - 1$. Since the primes 2 and 5 don't divide $10^\ell - 1$, this says they don't divide n either. Now that gives one half of the statement in (b).

We still need to convince ourselves that if $\gcd(n, 10) = 1$ then the decimal expansion of $1/n$ has the form $0.\overline{\mathtt{d}_1\mathtt{d}_2\mathtt{d}_3\ldots\mathtt{d}_\ell}$. Is that obviously true? Tracing through the steps above, we find this is the same as saying that if $\gcd(n, 10) = 1$ then there is exponent ℓ such that n divides $10^\ell - 1$... as it happens, this is true, and it's one of the great theorems of elementary number theory:

Theorem 1.2 (Euler's Theorem, Version 1). *Suppose $n \geqslant 3$ and $\gcd(n, 10) = 1$. There exists an integer $\ell \geqslant 1$ such that n divides $10^\ell - 1$.*

There's another way of stating this theorem that I find to be even more disarming, because anyone with a basic understanding of multiplication can appreciate its intrigue:

Theorem 1.3 (Euler's Theorem, Version 2). *Suppose we have a positive integer $n \geqslant 3$ whose last digit (or more precisely, whose digit in the ones place) is $1, 3, 7$, or 9. Among the list*

$$n, 2n, 3n, 4n, \ldots$$

of positive multiples of n, there is an integer of the form 999...9.

For instance, when $n = 13$, the list looks like

$$13, 26, 39, 52, 65, \ldots 999999 = 13 \cdot 76923, \ldots$$

and when $n = 41$ we find

$$41, 82, 123, 164, 205, \ldots 99999 = 41 \cdot 2439, \ldots$$

We can convince ourselves that Theorem 1.2 is what we need for the other half of statement (b) in Proposition 1.1, by considering the example of $n = 41$: Starting from $99999 = 10^5 - 1 = 41 \cdot 2439$, we find

$$\frac{1}{41} = \frac{2439}{10^5 - 1} = \frac{2439/10^5}{1 - 1/10^5} = \frac{2439}{10^5} + \frac{2439}{10^{10}} + \frac{2439}{10^{15}} + \cdots = 0.024390243902439 = 0.\overline{02439}.$$

In §1.2 and §1.4, we'll actually derive more precise versions of Euler's Theorem. But for now we ask: why is the version in Theorem 1.2 true? Let's be skeptical and suppose that it's *not* always true. That means we believe there is some value of n such that the infinite collection of integers

$$10 - 1, 10^2 - 1, 10^3 - 1, \ldots \tag{1.2}$$

doesn't contain any multiples of n. To state it differently, it means that when we divide any of these integers by n, the remainder is never 0. But what could that remainder possibly be? If it's not 0, then it must be $1, 2, 3, \ldots$ or $n-1$. But our list (1.2) is infinite, and there are only finitely many possible remainders. So that must mean that we can find (at least) two places in the list where the remainder is the same; let's call that common remainder r. Spelling this out, it means we have two exponents, say ℓ_1 and ℓ_2 with $1 \leqslant \ell_1 < \ell_2$, for which

$$10^{\ell_1} - 1 = nq_1 + r, \qquad 10^{\ell_2} - 1 = nq_2 + r.$$

(This is the formal way of saying that when $10^{\ell_i} - 1$ is divided by n, the quotient is q_i and the remainder is r.) When we subtract this expressions, we find

$$n(q_2 - q_1) = (10^{\ell_2} - 1) - (10^{\ell_1} - 1) = 10^{\ell_1}(10^{\ell_2 - \ell_1} - 1).$$

Now compare the left and right sides of this: we find that $10^{\ell_1}(10^{\ell_2-\ell_1} - 1)$ is a multiple of n. Since we assumed that n and 10 have no common factor, this must mean that in fact $10^{\ell_2-\ell_1} - 1$ is a multiple of n. But this goes against our belief that nothing in the list (1.2) is a multiple of n. So with this shaken belief, we decide to switch sides: Theorem 1.2 must always be true!

1.2 The period of $1/n$: A first approach

Up to this point, we've been a little loose with some of the terminology about our decimal expansions. For instance, one might argue that all numbers $1/n$ have an infinite decimal expansion, because we can write things such as

$$\frac{1}{2} = 0.5 = 0.50000... = 0.49999...$$

Similarly, one might be find the idea of the "repeating part" of an infinite decimal expansion to be ambiguous, since we can write things like

$$1/3 = 0.\overline{3} = 0.\overline{33} = 0.\overline{33333333}.$$

We can address these and other concerns by introducing the following device:

Definition 1.4. *For a real number* $\alpha \in [0, 1)$, *define the* shift operator $S : \mathbb{R} \to [0, 1)$ *by*

$$S(\alpha) = \{10\alpha\},$$

where $\{x\} = x - \lfloor x \rfloor$ *is the fractional part of the real number* x.

For example:

$$S(\pi) = \{10\pi\} = 10\pi - \lfloor 10\pi \rfloor = 10\pi - 31,$$

which in decimal expansions looks like

$$S(3.14159265...) = \{31.4159265...\} = 0.4159265...$$

As another example,

$$S\left(\frac{3}{7}\right) = \left\{\frac{30}{7}\right\} = \left\{4 + \frac{2}{7}\right\} = \frac{2}{7},$$

which in decimal expansions looks like

$$S(0.\overline{428571}) = \{4.\overline{285714}\} = 0.\overline{285714}.$$

These examples highlight why we're calling S the "shift" operator: it shifts the decimal point one place to the right and removes any digits to the left of the (newly located) decimal point. Now several concepts may be addressed in terms of this operator:

- A real number α has a finite decimal expansion if and only if $S^k(\alpha) = 0$ for some $k \geqslant 1$. (Note: $S^k(\alpha)$ denotes the composition of S with itself k times, and is not the kth power of $S(\alpha)$. For instance, $S^3(\alpha) = S(S(S(\alpha)))$.)
- Suppose that the decimal expansion of α is infinite. This expansion is periodic (i.e., "purely repeating") if $S^\ell(\alpha) = \alpha$ for some $\ell \geqslant 1$. The smallest such ℓ is called the *period* of the decimal expansion of α. We will often just call this the period of α.

n	**Decimal expansion of $1/n$**	**Smallest positive value of $10^\ell - 1$ divisible by n**
3	$0.\overline{3}$	$10^1 - 1 = 3 \cdot 3$
7	$0.\overline{142857}$	$10^6 - 1 = 7 \cdot 142857$
9	$0.\overline{1}$	$10^1 - 1 = 9 \cdot 1$
11	$0.\overline{09}$	$10^2 - 1 = 11 \cdot 9$
13	$0.\overline{076923}$	$10^6 - 1 = 13 \cdot 76923$
17	$0.\overline{0588235294117647}$	$10^{16} - 1 = 17 \cdot 5882352941176470$
19	$0.\overline{052631578947368421}$	$10^{18} - 1 = 19 \cdot 52631578947368421$
21	$0.\overline{047619}$	$10^6 - 1 = 21 \cdot 47619$
23	$0.\overline{0434782608695652173913}$	$10^{22} - 1 = 23 \cdot 434782608695652173913$
27	$0.\overline{037}$	$10^3 - 1 = 27 \cdot 37$
29	$0.\overline{0344827586206896551724137931}$	$10^{28} - 1 = 29 \cdot 344827586206896551724137931$

Table 1: An illustration of Proposition 1.6 for small values of n

- The decimal expansion of α is eventually periodic if $S^{h+\ell}(\alpha) = S^h(\alpha)$ for some $h \geqslant 0$ and $\ell \geqslant 1$. (In the case $h = 0$, we take $S^0(\alpha) = \alpha$; this allows periodic decimal expansions to be called eventually periodic as well.)

From this point forward, we're only going to concern ourselves with the case when the decimal expansion of $1/n$ is infinite and periodic. Thus: *From now on we assume that $n \geqslant 3$ is not divisible by 2 or 5.* Equivalently, we can state this assumption as $\gcd(n, 10) = 1$ or say that the last digit of n is either 1, 3, 7, or 9.

Our true focus of Part I is:

Question 1.5. *Given $n \geqslant 3$ with $\gcd(n, 10) = 1$, what is the period (of the decimal expansion) of $1/n$?*

As we saw in §1.1 (see equation (1.1) above), the use of geometric series and Theorem 1.2 allows us to relate the period of $1/n$ to a certain divisibility statement:

Proposition 1.6. *Suppose $n \geqslant 3$ and $\gcd(n, 10) = 1$. The period of $1/n$ is equal to the smallest integer $\ell \geqslant 1$ such that $10^\ell - 1$ is divisible by n.*

In Table 1, we've collected a small sample of data that displays the correspondence in this proposition. As the wild variation in this table suggests, for a given value of n, it's not so clear what the period of $1/n$ will be without doing a lot of calculation. For instance, just to select a rather large and arbitrary n, do you have a reasonable guess for the period of $1/15247$?[2]

But the situation is not quite as hopeless as Table 1 suggests. To see what we can say, let's start by making a careful study of the example $1/21 = 0.\overline{047619}$. What happens if we apply the shift operator to this number several times? In decimals, the outcome can be displayed as follows:

$$0.\overline{047619} \overset{S}{\rightsquigarrow} 0.\overline{476190} \overset{S}{\rightsquigarrow} 0.\overline{761904} \overset{S}{\rightsquigarrow} 0.\overline{619047} \overset{S}{\rightsquigarrow} 0.\overline{190476} \overset{S}{\rightsquigarrow} 0.\overline{904761} \overset{S}{\rightsquigarrow} 0.\overline{047619}$$

Alternatively (remembering that the definition of S is not actually given in terms of decimal expansions), we can also display this as

$$\frac{1}{21} \overset{S}{\rightsquigarrow} \frac{10}{21} \overset{S}{\rightsquigarrow} \frac{16}{21} \overset{S}{\rightsquigarrow} \frac{13}{21} \overset{S}{\rightsquigarrow} \frac{4}{21} \overset{S}{\rightsquigarrow} \frac{19}{21} \overset{S}{\rightsquigarrow} \frac{1}{21}$$

If we look at all of the fractions produced in the sequence above, they're all of the form $a/21$ where the numerator a has two properties:

[2] The period of $1/15427$ happens to be 2496. See what I mean?

n	2	3	4	5	6	7	8	9	10	11	12	13	14	15	16	17	18	19	20
$\phi(n)$	1	2	2	4	2	6	4	6	4	10	4	12	6	8	8	16	6	18	8

Table 2: Small values of Euler's ϕ-function

1. $1 \leqslant a < 21$
2. There are no primes dividing both a and 21. Equivalently, $\gcd(a, 21) = 1$.

Is this sort of thing going to be true for $1/n$ in general? Yes, and it follows from our assumption that n is not divisible by 2 or 5. To start with, if we define the integer $c = \lfloor 10/n \rfloor$ then

$$S\left(\frac{1}{n}\right) = \frac{10}{n} - \left\lfloor \frac{10}{n} \right\rfloor = \frac{10}{n} - c = \frac{10 - cn}{n}.$$

Is the fraction $(10 - cn)/n$ written in reduced form? If not, there is a prime p dividing both n and $10 - cn$ that we'll be able to cancel from the numerator and denominator; and this means p will also divide $10 = (10 - cn) - cn$, since this is a difference of two multiples of p. But we assumed n and 10 have no common prime divisors, so this must mean that no such p exists. In other words, $(10 - cn)/n$ is indeed in reduced form, and so we can say that $S(1/n) = a/n$ where $\gcd(a, n) = 1$. Moreover, we'll have $1 \leqslant a < n$ since $S(1/n) \in (0, 1)$.

With each successive application of S, we'll be able to make a similar argument: the fact that $\gcd(n, 10) = \gcd(a, n) = 1$ can be used to show $S(a/n) = a'/n$ for some integer a' satisfying (1) $1 \leqslant a' < n$ and (2) $\gcd(a', n) = 1$. Looking at these two properties that we've highlighted, it's now worth introducing the following number-theoretic function:

Definition 1.7. *Let $m \geqslant 2$ be a positive integer.* Euler's ϕ-function *is defined to be the number $\phi(m)$ of integers a that satisfy $1 \leqslant a < m$ and $\gcd(a, m) = 1$. Equivalently, $\phi(m)$ is equal to the number of fractions in the list*

$$\frac{1}{m}, \frac{2}{m}, \ldots, \frac{m-1}{m}$$

whose reduced form looks like a/m.

For instance, among the integers $1, 2, 3, \ldots, 11$, only $1, 5, 7$, and 11 don't have any prime factors in common with 12; hence $\phi(12) = 4$. Equivalently, if we look at the fractions $1/12, 2/12, 3/12 \ldots, 11/12$, there are exactly 4 in this list that can't be simplified, namely $1/12, 5/12, 7/12$, and $11/12$. As another example, none of the integers $1, 2, 3, 4, 5, 6$ are divisible by the prime 7, hence all 6 of the fractions $1/7, 2/7, \ldots, 6/7$ are already in reduced form, and so $\phi(7) = 6$. Table 2 lists the first few values of $\phi(n)$, and gives the impression that this function behaves somewhat unpredictably. But, as we'll discuss in §1.5, this is due to the fact that $\phi(n)$ is closely tied to the prime factorization of n.

Let's summarize our discussion of the period of $1/n$ up to this point, by introducing some convenient notation: For a real number α, let

$$T_\alpha = \{S^k(\alpha) \mid k \geqslant 1\} \tag{1.3}$$

be the set of all real numbers that can possibly be obtained from α by repeated application of S. Also define

$$U_n = \left\{ \frac{a}{n} \;\middle|\; 1 \leqslant a < n \text{ and } \gcd(a, n) = 1 \right\}. \tag{1.4}$$

Then by assuming that the expansion of $1/n$ is periodic with period ℓ, we can say that

$$T_{1/n} = \left\{ S\left(\frac{1}{n}\right), S^2\left(\frac{1}{n}\right), \ldots, S^\ell\left(\frac{1}{n}\right) \right\}$$

and that $T_{1/n} \subseteq U_n$. Now look at the sizes of these two sets. Of course $\#T_{1/n} \leqslant \ell$, but in fact it can't be any smaller than this: if it were, we would have $S^{k_1}(1/n) = S^{k_2}(1/n)$ for some $1 \leqslant k_1 < k_2 \leqslant \ell$, and this would give

$$S^{\ell-(k_2-k_1)}\left(\frac{1}{n}\right) = S^{\ell-k_2}\left(S^{k_1}\left(\frac{1}{n}\right)\right) = S^{\ell-k_2}\left(S^{k_2}\left(\frac{1}{n}\right)\right) = S^{\ell}\left(\frac{1}{n}\right) = \frac{1}{n},$$

contradicting the fact that the period of $1/n$ is ℓ. So we have $\#T_n = \ell$. Moreover, comparing the definitions of $\phi(n)$ and U_n, we see that $\#U_n = \phi(n)$. So the inclusion $T_{1/n} \subseteq U_n$ implies:

Proposition 1.8. *If* $\gcd(n, 10) = 1$, *then the period of the decimal expansion of* $1/n$ *is at most* $\phi(n)$.

As a byproduct of this analysis, we can combine this proposition with Proposition 1.6 to obtain the following improvement of Theorem 1.2:

Theorem 1.9 (Euler's Theorem, Version 3). *Suppose* $n \geqslant 3$ *and* $\gcd(n, 10) = 1$. *There exists an integer* $1 \leqslant \ell \leqslant \phi(n)$ *such that* n *divides* $10^{\ell} - 1$.

While it won't be our final word about the period of $1/n$, let's not overlook the fact that Proposition 1.8 already tells us something nontrivial. Indeed, looking at Table 1, it may not be clear that the period of $1/n$ is limited in any way by the size of n. But the definition of $\phi(n)$ implies that $\phi(n) \leqslant n - 1$, and so the period of $1/n$ is at most $n - 1$ as well.

1.3 Interlude: Magic and mystery

Those values of n for which the period of $1/n$ is $n - 1$ should be regarded as special, because their decimal expansions are "as large as possible." The first few values of n for which this occurs are $n = 7, 17, 19, 23, \ldots$. In general, we can only have $\phi(n) = n - 1$ if n is prime, a fact which follows from the definition of $\phi(n)$. But as the examples of $n = 3, 11$, and 13 show, knowing that n is prime doesn't guarantee that the period of $1/n$ is $n - 1$. So we ask:

Question 1.10. *For which primes* $p \neq 2, 5$ *is the period of* $1/p$ *equal to* $p - 1$?

By dwelling upon these special kinds of primes, we'll come face-to-face with a little bit of magic and a great mystery. First for the magic: we'll look at the case of $p = 7$. Take the expansion $1/7 = 0.\overline{142857}$ and consider the associated integer $m_7 = 142857$. Now look at the first few multiples of m_7:

$$\begin{aligned} m_7 &= 142857 \\ 2m_7 &= 285714 \\ 3m_7 &= 428571 \\ 4m_7 &= 571428 \\ 5m_7 &= 714285 \\ 6m_7 &= 857142 \end{aligned}$$

If you compare the sequence of decimal digits in these 6 multiples, you see they're just cyclic permutations of one another. (That is, we can get from one to any other by removing some initial sequence of digits on the left and reattaching it to the right side.) Another example of this phenomenon comes from $1/17 = 0.\overline{0588235294117647}$. Putting $m_{17} = 588235294117647$, its first 16 multiples are

0588235294117647, 1176470588235294, 1764705882352941, 2352941176470588,
2941176470588235, 3529411764705882, 4117647058823529, 4705882352941176,
5294117647058823, 5882352941176470, 6470588235294117, 7058823529411764,
7647058823529411, 8235294117647058, 8823529411764705, 9411764705882352.

Numbers such as m_7 and m_{17} are called *cyclic numbers*; in general a positive integer is a d-digit cyclic number if the decimal representations of its first $d-1$ multiples are cyclic permutations of the integer itself. (As with the example m_{17} above, this might require adding some copies of the digit 0 to the left side of the integer.) As you can guess, cyclic numbers are quite rare! Just as with Theorem 1.3, it's worth noting that these intriguing numbers may be appreciated even by young students.

If we assume the decimal expansion of $1/p$ has period p, why does the expansion yield a cyclic number in this way? It goes back to the fact that, under this assumption, we have $U_p = T_{1/p}$, i.e., any fraction a/p with $1 \leqslant a < p$ is equal to $S^k(1/p)$ for some $1 \leqslant k \leqslant p-1$. Thus

$$\frac{1}{p} = 0.\overline{\mathtt{d}_1\mathtt{d}_2\mathtt{d}_3\ldots\mathtt{d}_{p-1}} \quad\Longrightarrow\quad \frac{a}{p} = S^k\left(\frac{1}{p}\right) = 0.\overline{\mathtt{d}_{k+1}\ldots\mathtt{d}_{p-1}\mathtt{d}_1\ldots\mathtt{d}_k}$$

But arguing with geometric series as we did in (1.1), we can let m_p be the positive integer whose decimal representation is $\mathtt{d}_1\mathtt{d}_2\mathtt{d}_3\ldots\mathtt{d}_{p-1}$. Then

$$\frac{1}{p} = 0.\overline{\mathtt{d}_1\mathtt{d}_2\mathtt{d}_3\ldots\mathtt{d}_{p-1}} = \frac{m_p}{10^{p-1}-1} \quad\Longrightarrow\quad \frac{a}{p} = \frac{am_p}{10^{p-1}-1}$$

This means that

$$0.\overline{\mathtt{d}_{k+1}\ldots\mathtt{d}_{p-1}\mathtt{d}_1\ldots\mathtt{d}_k} = \frac{am_p/10^{p-1}}{1-1/10^{p-1}} = \frac{am_p}{10^{p-1}} + \frac{am_p}{10^{2(p-1)}} + \frac{am_p}{10^{3(p-1)}} + \cdots$$

Since $a < p$, we have $am_p < pm_p = 10^{p-1}-1 < 10^{p-1}$, we conclude from this that the decimal representation of am_p is $\mathtt{d}_{k+1}\ldots\mathtt{d}_{p-1}\mathtt{d}_1\ldots\mathtt{d}_k$.

In addition to cyclic numbers, there is another motivation for looking at primes p for which $1/p$ has period $p-1$. According to Proposition 1.6, when $1/p$ has period $p-1$, this means that the none of the integers

$$10-1, 10^2-1, 10^3-1, \ldots, 10^{p-2}-1$$

are divisible by p. In the language of number theory, this says that 10 is a *primitive root* of p. In general, we have the following definition:

Definition 1.11. *Let p be prime and b an integer that is not a multiple of p. We say that b is a* primitive root *of p if none of the integers*

$$b-1, b^2-1, b^3-1, \ldots, b^{p-2}-1$$

are divisible by p.

When one first encounters the idea of primitive roots in number theory, it takes awhile to digest it. In particular, it seems unclear why anyone would care about such a concept at all. But it turns out to be a big deal that, for a given prime p, we can always find at least one integer b which is a primitive root of p, because this fact makes it much easier to prove certain theorems about primes. It may also be surprising to hear that primitive roots have found uses outside of number theory: they're commonly used in cryptography, especially in the *ElGamal cryptosystem* [HPS], that we rely upon for internet security and privacy, and they've even found an application in radar and sonar technology [Sil].

Turning back to Question 1.10, let's step back and look at some data. Figure 1 gives a visual representation of the period of $1/p$ for the first 100 primes. Let p_k denote the kth prime number (so $p_1 = 2, p_2 = 3, p_3 = 5, \ldots$). In Figure 1, the points

$$\left(k, \frac{\text{period of } 1/p_k}{p_k - 1}\right)$$

are displayed for $1 \leqslant k \leqslant 100$, so that in particular the y-coordinate is always at most 1. (In the figure you might also notice the anomalous values at $p_1 = 2$ and $p_3 = 5$; we're not supposed to consider these, so I've just set the value to 0.) Do you see a pattern in this graph? I sure don't! This is a mysterious sequence, and

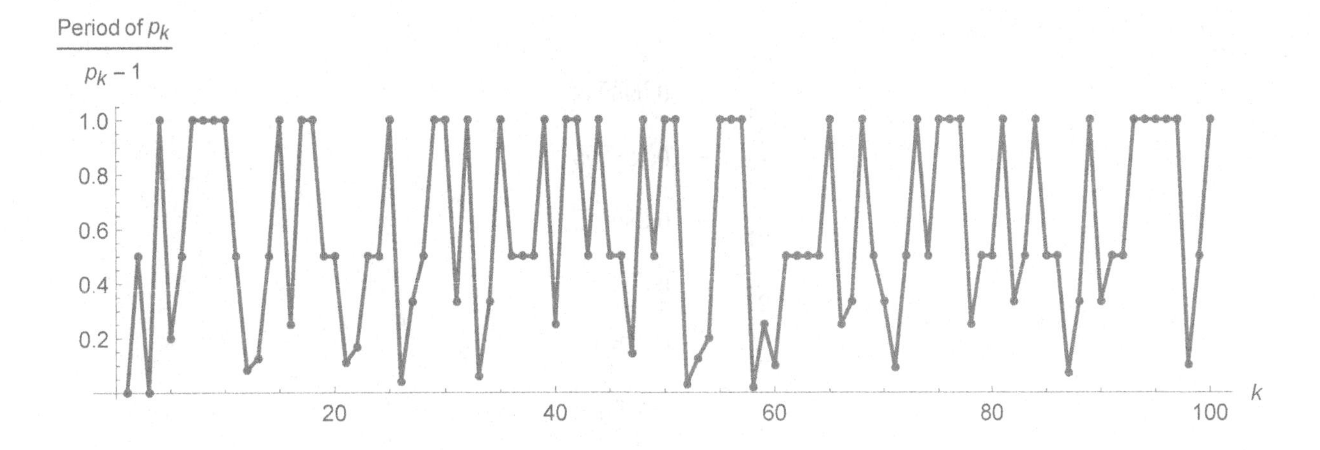

Figure 1: $\left(\frac{\text{period of } 1/p_k}{p_k - 1}\right)$ versus k

it looks like it doesn't behave in any predictable way. For instance, since you see the first 100 terms of the sequence displayed in the graph, can you use this to confidently predict what 101st term in the sequence is? I can't!

As an indication of just how little we understand the behavior of this graph, mathematicians have not even answered a question of Gauss, which asks whether there will always be points lying on the line $y = 1$ as $k \to \infty$; note that these points correspond to primes p_k for which $1/p_k$ has the maximum period $p_k - 1$. In the early twentieth century, Emil Artin gave a conjecture that, if true, would answer Gauss' question in the affirmative, and would even tell us "how often" $1/p$ has period $p - 1$:

Conjecture 1.12 (Artin). *For a large "random" prime p, the probability that the decimal expansion of* $1/p$ *will have period p is about* 37%. *More precisely, when one looks at all primes* $p \leqslant N$, *the proportion of those for which* $1/p$ *has period* $p - 1$ *approaches*

$$\prod_{q \text{ prime}} \left(1 - \frac{1}{q(q-1)}\right) = 0.3739558136...$$

as $N \to \infty$.

In any case, we see that Question 1.10 is not so innocent as it might first seem. It shows how these everyday decimal expansions are rooted in unexpectedly mysterious and uncharted terrain.

1.4 The period of $1/n$: A refinement

Now let's go back to the more general considerations we had at the end of §1.2. As in that section, we have an integer $n \geqslant 3$ such that $\gcd(n, 10) = 1$, and we let ℓ be the period of the decimal expansion of $1/n$. To refine the result in Proposition 1.8, let's revisit the inclusion $T_{1/n} \subseteq U_n$ that we had shown. If $\ell < \phi(n)$, then there are fractions a/n inside U_n that cannot be reached by repeatedly applying the shift operator to $1/n$. So we're going to start by taking a closer look at the general elements of U_n.

Let's revisit the example of $n = 21$ again. We have

$$T_{1/21} = \left\{\frac{1}{21}, \frac{4}{21}, \frac{10}{21}, \frac{13}{21}, \frac{16}{21}, \frac{19}{21}\right\} \subseteq \left\{\frac{1}{21}, \frac{2}{21}, \frac{4}{21}, \frac{5}{21}, \frac{8}{21}, \frac{10}{21}, \frac{11}{21}, \frac{13}{21}, \frac{16}{21}, \frac{17}{21}, \frac{19}{21}, \frac{20}{21}\right\} = U_{21}$$

so the six elements in U_{21} that are not of the form $S^k(1/21)$ are

$$\left\{\frac{2}{21}, \frac{5}{21}, \frac{8}{21}, \frac{11}{21}, \frac{17}{21}, \frac{20}{21}\right\}.$$

Let's take a look at their decimal expansions:

$$\begin{aligned}
\frac{2}{21} &= 0.\overline{095238} \\
\frac{5}{21} &= 0.\overline{238095} \\
\frac{8}{21} &= 0.\overline{380952} \\
\frac{11}{21} &= 0.\overline{523809} \\
\frac{17}{21} &= 0.\overline{809523} \\
\frac{20}{21} &= 0.\overline{952380}.
\end{aligned}$$

We can see that these six fractions are all related by the shift operator S! So the following picture emerges. When we unleash S upon the 12-element set U_{21}, by applying it to its elements again and again, we wind up with a natural division of U_{21} into two 6-element subsets, namely T_{21} and its complement. Both of these subsets have the following property: within each subset, we can use S (repeatedly) to turn any one element into any other.

So what kind of picture will we get if we unleash S upon U_n for more general values of n? Let's first recall from (1.3) that the set $T_{1/n}$ is one instance of a more general construction. Fixing any element a/n inside U_n, let's now ask what can be said about $T_{a/n} = \{S^k(a/n) \mid k \geqslant 1\}$. First of all, I claim that the fraction a/n is periodic, and that its period is no larger than ℓ, the period of $1/n$. To see this, let's recall from (1.1) that if $1/n = 0.\overline{\mathtt{d}_1\mathtt{d}_2\ldots\mathtt{d}_\ell}$ and we put $m = \mathtt{d}_1 \cdot 10^{\ell-1} + \ldots + \mathtt{d}_{\ell-1} \cdot 10 + \mathtt{d}_\ell$, then the geometric series identity allows us to conclude that $1/n = m/(10^\ell - 1)$, and that n does not divide $10^k - 1$ for any $1 \leqslant k < \ell$. We're going to reverse this process to learn about the decimal expansion of a/n. Since $1 \leqslant a < n$, we can say that $am < nm = 10^\ell - 1 < 10^\ell$. So the decimal representation of am will have the form

$$am = \mathtt{e}_1 \cdot 10^{\ell-1} + \mathtt{e}_2 \cdot 10^{\ell-2} + \cdots + \mathtt{e}_{\ell-1} \cdot 10 + \mathtt{e}_\ell,$$

which we may write more succinctly as $am = \mathtt{e}_1\mathtt{e}_2\ldots\mathtt{e}_\ell$. (Just as with $\mathtt{d}_1, \mathtt{d}_2, \ldots$, some of the leading digits $\mathtt{e}_1, \mathtt{e}_2\ldots$, might be 0.) Thus we have

$$\begin{aligned}
\frac{a}{n} &= \frac{am}{nm} \\
&= \frac{am}{10^\ell} \cdot \frac{1}{(1 - 1/10^\ell)} \\
&= \left(\frac{\mathtt{e}_1}{10} + \frac{\mathtt{e}_2}{10^2} + \cdots + \frac{\mathtt{e}_{\ell-1}}{10^{\ell-1}} + \frac{\mathtt{e}_\ell}{10^\ell}\right) \cdot \left(1 + \frac{1}{10^\ell} + \frac{1}{10^{2\ell}} + \cdots\right) \\
&= \frac{\mathtt{e}_1}{10} + \frac{\mathtt{e}_2}{10^2} + \cdots + \frac{\mathtt{e}_\ell}{10^\ell} + \frac{\mathtt{e}_1}{10^{\ell+1}} + \frac{\mathtt{e}_2}{10^{\ell+2}} + \cdots + \frac{\mathtt{e}_\ell}{10^{2\ell}} + \cdots \\
&= 0.\overline{\mathtt{e}_1\mathtt{e}_2\ldots\mathtt{e}_\ell}.
\end{aligned}$$

So indeed the period of a/n is at most ℓ. In terms of the shift operator, we can write $S^\ell(a/n) = a/n$.

Could the period of a/n be less than ℓ? If it had period $k < \ell$, that would mean (by using the same kind of geometric series argument we just used) that

$$\frac{a}{n} = \frac{m'}{10^k - 1}$$

for some integer $1 \leqslant m' < 10^k - 1$. Cross-multiplying gives $nm' = a(10^k - 1)$, so in particular n divides $a(10^k - 1)$. But by definition of U_n, a and n do not share any prime factors, and so in fact n divides $10^k - 1$. But this is a problem, because we already stated that ℓ is the smallest exponent for which n divides $10^\ell - 1$. The conclusion is therefore that the period a/n must equal ℓ.

So now we can say that, for any a/n in U_n, the set $T_{a/n}$ can be described more simply as

$$T_{a/n} = \left\{ S\left(\frac{a}{n}\right), S^2\left(\frac{a}{n}\right), \ldots, S^\ell\left(\frac{a}{n}\right) \right\}.$$

Just as we did toward the end of §1.2 for the set $T_{1/n}$, we can conclude that this set contains ℓ distinct elements (i.e., that $S^{k_1}(a/n) \neq S^{k_2}(a/n)$ whenever $1 \leqslant k_1 < k_2 \leqslant \ell$).

Here's the picture we have of U_n so far: When we apply S repeatedly to any element a/n of U_n, we generate an ℓ-element subset $T_{a/n}$. If we take two distinct elements a_1/n and a_2/n, how much do the sets $T_{a_1/n}$ and $T_{a_2/n}$ have in common? In the example of $n = 21$, we saw that were the only two extremes: Either they were actually equal (e.g., $T_{1/21} = T_{13/21}$ and $T_{2/21} = T_{8/21}$) or they were disjoint (e.g., $T_{1/n} \cap T_{2/n} = \varnothing$). And this turns out to happen in the general case as well. To see this, let's suppose that $T_{a_1/n}$ and $T_{a_2/n}$ have at least one common element, say b/n. Then for some exponents $1 \leqslant k_1, k_2 \leqslant \ell$, we have

$$S^{k_1}\left(\frac{a_1}{n}\right) = \frac{b}{n} = S^{k_2}\left(\frac{a_2}{n}\right).$$

Now apply S multiple times to this, using the fact that a_2/n has period ℓ:

$$S^{k_1+\ell-k_2}\left(\frac{a_1}{n}\right) = S^{\ell-k_2}\left(S^{k_1}\left(\frac{a_1}{n}\right)\right) = S^{\ell-k_2}\left(S^{k_2}\left(\frac{a_2}{n}\right)\right) = S^{\ell}\left(\frac{a_2}{n}\right) = \frac{a_2}{n}.$$

This shows us that a_2/n belongs to $T_{a_1/n}$, and so anything we can generate by repeatedly applying S to a_2/n can in fact be obtained by applying S to a_1/n. So we conclude that $T_{a_2/n} \subseteq T_{a_1/n}$, and a symmetric argument shows the opposite inclusion. Hence there are only two possibilities: $T_{a_1/n} \cap T_{a_2/n} = \varnothing$ or $T_{a_1/n} = T_{a_2/n}$.

So our final picture of U_n has been painted! It is divided into of a number of subsets—let's say there are k of them—that (i) each have ℓ elements and (ii) do not intersect each other. We conclude that the number of elements in U_n is equal to $k\ell$. But we also recall that $\#U_n = \phi(n)$, and this brings us to the following refinement of Proposition 1.8:

Theorem 1.13. *If* $\gcd(n, 10) = 1$, *then the period of the decimal expansion of* $1/n$ *divides* $\phi(n)$.

Notice that while it doesn't describe the period of $1/n$ completely, Theorem 1.13 gives us much more information than Proposition 1.8. For instance, consider what each one tells us about the period ℓ of $1/101$. Since 101 is prime, we have $\phi(101) = 100$ and so Proposition 1.8 tells us that ℓ is something between 1 and 100. But Theorem 1.13 tells us that ℓ actually divides 100, so ℓ is one of the numbers 1, 2, 4, 5, 10, 20, 25, 50, or 100. This is a much smaller list of possibilities!

To finish this section, let's return to Euler's Theorem. From Proposition 1.6, Theorem 1.13 is equivalent to saying that, when $\gcd(n, 10) = 1$, there exists an exponent $\ell \geqslant 1$ such that n divides $10^\ell - 1$, and furthermore ℓ is a divisor of $\phi(n)$. So we may put $\phi(n) = k\ell$ and conclude[3] that

$$10^{\phi(n)} - 1 = (10^\ell)^k - 1 = (10^\ell - 1)\left((10^\ell)^{k-1} + (10^\ell)^{k-2} + 10^\ell + \cdots + 1\right)$$

is also a multiple of n. So we arrive at a more faithful version of Euler's fundamental result:

Theorem 1.14 (Euler's Theorem, Version 4)**.** *For any positive integer* n *such that* $\gcd(n, 10) = 1$, *the integer* $10^{\phi(n)} - 1$ *is a multiple of* n.

For a prime $p \neq 2, 5$ we have $\phi(p) = p - 1$, and Theorem 1.14 gives the following fascinating elementary statement (originally due to Fermat): the integer

$$\underbrace{9999\ldots9}_{p-1 \text{ times}}$$

is divisible by p.

[3]This is just an application of the usual formula for the sum of a *finite* geometric series: When $a \neq 1$,

$$1 + a + a^2 + \cdots + a^{k-2} + a^{k-1} = \frac{a^k - 1}{a - 1}.$$

1.5 The period of $1/n$: The heart of the matter

This goal of this section is to arrive at the foundations of Question 1.5. If we're handed the integer n and we want to determine the period of the decimal expansion of $1/n$, how easy can we make this task if we want to avoid computing the actual decimal expansion?

To get to an answer, we're going to need the following fact:

Proposition 1.15. *Suppose that* $\gcd(n, 10) = 1$ *and let* ℓ *be the period of* $1/n$. *If* n *divides* $10^k - 1$ *for some* $k \geqslant 1$, *then* ℓ *divides* k.

The key to showing this is the geometric series connection that we've used several times by now. Assuming $10^k - 1 = nq$, we may write the decimal representation of q in the form $\mathtt{e}_1\mathtt{e}_2\ldots\mathtt{e}_k$ and then find

$$\frac{1}{n} = \frac{q}{10^k - 1} = \frac{q/10^k}{1 - 1/10^k} = \frac{q}{10^k} + \frac{q}{10^{2k}} + \frac{q}{10^{3k}} + \cdots = 0.\overline{\mathtt{e}_1\mathtt{e}_2\ldots\mathtt{e}_k}.$$

This shows that the decimal expansion of $1/n$ repeats the k-digit sequence $\mathtt{e}_1, \mathtt{e}_2, \ldots, \mathtt{e}_k$. But the period of $1/n$ is ℓ, and so if we write $1/n = 0.\overline{\mathtt{d}_1\mathtt{d}_2\ldots\mathtt{d}_\ell}$, the only way these two decimal expansions of $1/n$ can coexist is if k is a multiple of ℓ and the sequence $\mathtt{e}_1, \mathtt{e}_2, \ldots, \mathtt{e}_k$ just consists of k/ℓ repetitions of the smaller sequence $\mathtt{d}_1, \mathtt{d}_2, \ldots, \mathtt{d}_\ell$. Proposition 1.15 follows from this.[4]

The next result tells us that when n_1 and n_2 don't share any prime factors, we can easily figure out the period of $1/n_1n_2$ once we know the periods of $1/n_1$ and $1/n_2$:

Theorem 1.16. *Suppose* $\gcd(n_1, 10) = \gcd(n_2, 10) = 1$, *let* $1/n_1$ *have period* ℓ_1, *and* $1/n_2$ *have period* ℓ_2. *If* $\gcd(n_1, n_2) = 1$, *then the period of* $1/n_1n_2$ *is* $\operatorname{lcm}(\ell_1, \ell_2)$, *the least common multiple of* ℓ_1 *and* ℓ_2.

This theorem allows us to use a small set of facts to build up more impressive ones. For instance, let's start with the quantities

$$\frac{1}{7} = 0.\overline{142857}, \quad \frac{1}{27} = 0.\overline{037}, \quad \frac{1}{41} = 0.\overline{02439},$$

which involve relatively small denominators. Since $189 = 7 \cdot 27$ and $\gcd(7, 27) = 1$, this means that the period of $1/189$ is $\operatorname{lcm}(6, 3) = 6$; building on this, we have $7749 = 189 \cdot 41$ and $\gcd(41, 189) = 1$, so the period of $1/7749$ is $\operatorname{lcm}(6, 5) = 30$. Keep in mind we didn't do the work to calculate the actual decimal expansion of $1/7749$ but, whatever it is, it will have a period of 30! We should also note that the theorem has its limitations; for instance, you can't start from $1/3 = 0.\overline{3}$ and $1/27 = 0.\overline{0.037}$ and conclude $1/81$ has period $\operatorname{lcm}(1, 3) = 3$, because $\gcd(3, 27) \neq 1$. (In fact the period of $1/81$ is 9.)

Now let's convince ourselves that Theorem 1.16 is true. We already have ℓ_1 as the period of $1/n_1$ and ℓ_2 as the period of $1/n_2$, so let's define ℓ_3 to be the period of $1/n_1n_2$. To show that $\ell_3 = \operatorname{lcm}(\ell_1, \ell_2)$, we're going to show that these quantities both divide each other.

First we'll show that ℓ_3 divides $\operatorname{lcm}(\ell_1, \ell_2)$. By its very definition, $\operatorname{lcm}(\ell_1, \ell_2)$ is a multiple of ℓ_1; let's say $\operatorname{lcm}(\ell_1, \ell_2) = \ell_1 c$. Since

$$10^{\operatorname{lcm}(\ell_1,\ell_2)} - 1 = (10^{\ell_1})^c - 1 = (10^{\ell_1} - 1)\Big((10^{\ell_1})^{c-1} + (10^{\ell_1})^{c-2} + \cdots + 1\Big),$$

we can conclude that n_1 divides $10^{\operatorname{lcm}(\ell_1,\ell_2)} - 1$, since it divides $10^{\ell_1} - 1$. By a completely similar argument we can conclude n_2 also divides $10^{\operatorname{lcm}(\ell_1,\ell_2)} - 1$. Now let's use the fact that $\gcd(n_1, n_2) = 1$: since n_1 and n_2 have no prime factors in common, the fact that they both divide the same integer means that their product must also divide that integer. (This is perhaps most intuitively seen by considering prime factorizations.)

[4]This argument appeals to your intuition about repeating decimal expansions, particularly when I used the words "the only way." If this feels a little slippery to you, the proposition can also be proved without resorting to decimal expansions. One can instead use the Division Algorithm (another cornerstone of elementary number theory!) to write $k = \ell q + r$, where $0 \leqslant r < \ell$, show that $10^r - 1$ is a multiple of n, and apply what we know about ℓ from Proposition 1.6.

Hence n_1n_2 divides $10^{\text{lcm}(\ell_1,\ell_2)}-1$, and now we can apply our new tool, Proposition 1.15, to conclude that the period ℓ_3 of $1/n_1n_2$ divides $\text{lcm}(\ell_1,\ell_2)$.

To show that $\text{lcm}(\ell_1,\ell_2)$ divides ℓ_3, we're going to wield Proposition 1.15 again. Since n_1n_2 divides $10^{\ell_3}-1$, this of course implies that n_1 divides $10^{\ell_3}-1$, and so we conclude that ℓ_1 divides ℓ_3. We use the same argument to show that ℓ_2 divides ℓ_3 as well. Since both ℓ_1 and ℓ_2 divide ℓ_3, it follows that $\text{lcm}(\ell_1,\ell_2)$ must also divide ℓ_3. (This last step exhibits a key property about least common multiples and, like the fact in the previous paragraph, may again be seen using prime factorizations.) So this gives our desired equality $\ell_3 = \text{lcm}(\ell_1,\ell_2)$.

Now let's discuss Euler's ϕ-function a little more. We've already noted that, by the definitions of prime and composite, we have $\phi(n) = n-1$ if and only if n is prime. What if n is a power of a prime? It's not too much harder to compute $\phi(p^k)$ directly from Definition 1.7. First we write out all of the numbers

$$1, 2, 3, \ldots, p^k - 1$$

and ask which ones have any prime factors in common with p^k. But the only prime factor of p^k is p, so we're really just asking how many multiples of p there are in this list. If we write out all of these the multiples, we get

$$p, 2p, 3p, \ldots, p^k - p = p(p^{k-1}-1).$$

So there are $p^{k-1}-1$ multiples of p here, and after crossing them all out from the first list we're left with

$$(p^k-1)-(p^{k-1}-1) = p^{k-1}(p-1)$$

numbers. Therefore

$$\phi(p^k) = p^{k-1}(p-1). \tag{1.5}$$

We can use this, for instance, along with Theorem 1.13 to say that the period of $1/7^8$ is a divisor of $7^7\cdot(7-1) = 4941258$ (a fact which gives us 32 possibilities for this period).

Let me now tell you how we can calculate $\phi(n)$ for any positive integer n. In our quest for the most precise information about the period of $1/n$, we won't actually need to use this formula, but I would feel bad to leave it out after coming this far. A key property of the ϕ-function is that it is *multiplicative*, which means that

$$\phi(n_1n_2) = \phi(n_1)\phi(n_2) \quad \text{if } \gcd(n_1,n_2) = 1.$$

Since we won't ultimately use this result, I won't try to justify this multiplicative property; but most textbooks on elementary number theory will have a proof. Now imagine that you have some integer n and you know its prime factorization. For concreteness, let's write it as

$$n = p_1^{h_1}p_2^{h_2}\cdots p_r^{h_r},$$

where $r \geqslant 1$, each p_i is prime, each $h_i \geqslant 1$, and $p_1 < p_2 < \ldots < p_r$. Since $\gcd(p_1^{h_1}, p_2^{h_2}) = 1$, we have

$$\phi(p_1^{h_1}p_2^{h_2}) = \phi(p_1^{h_1})\phi(p_2^{h_2}) = p_1^{h_1-1}(p_1-1)p_2^{h_2-1}(p_2-1).$$

We also have $\gcd(p_1^{h_1}p_2^{h_2}, p_3^{h_3}) = 1$, so

$$\phi(p_1^{h_1}p_2^{h_2}p_3^{h_3}) = \phi(p_1^{h_1}p_2^{h_2})\phi(p_3^{h_3}) = p_1^{h_1-1}(p_1-1)p_2^{h_2-1}(p_2-1)p_3^{h_3-1}(p_3-1).$$

Continuing in this fashion, we arrive at the formula

$$\phi(n) = p_1^{h_1-1}(p_1-1)p_2^{h_2-1}(p_2-1)\cdots p_r^{h_r-1}(p_r-1). \tag{1.6}$$

Let's take a look what this formula says about the period of $1/31941$. Its prime factorization is $31941 = 3^3\cdot 7\cdot 13$, and so (1.6) gives

$$\phi(31941) = 3^2\cdot(3-1)\cdot 7^0\cdot(7-1)\cdot 13^1\cdot(13-1) = 16848.$$

(This is much better than trying to use Definition 1.7 to calculate $\phi(31941)$!) What does Theorem 1.13 then tell us about the period of 1/31941? It divides $\phi(31941) = 16848$, which ends up giving us 50 possibilities; in particular, without any further information, the period could be as large as 16848.

With a little more finesse, though, we can actually do better than this. Let's suppose we don't know the periods of $1/3^3 = 1/27$, $1/7$ and $1/13^2 = 1/169$, and just call them ℓ_{27}, ℓ_7, and ℓ_{169}. Using only the simple formula (1.5) and Theorem 1.13, we can say that ℓ_{27} divides $\phi(27) = 18$, ℓ_7 divides $\phi(7) = 6$, and ℓ_{169} divides $\phi(169) = 156$. Taking the first two of these statements, we can say that $\mathrm{lcm}(\ell_{27}, \ell_7)$ must divide $\mathrm{lcm}(18, 6) = 18$; so by Theorem 1.16, the period ℓ_{189} of $1/(27 \cdot 7) = 1/189$ divides 18. Then, as $\gcd(189, 169) = 1$, we use Theorem 1.16 once more to conclude that the period of $1/(189 \cdot 169) = 1/31941$, which equals $\mathrm{lcm}(\ell_{189}, \ell_{169})$, must divide $\mathrm{lcm}(18, 156) = 468$. This cuts down the number of possibilities for the period of 1/31941 to just 18, and implies it is no larger than 468—a bound that's 26 times smaller than the one obtained in the previous paragraph!

This example is an illustration of the way in which the period of $1/n$ is determined by the periods of the numbers $1/p^h$, for the prime powers p^h appearing in its prime factorization. Here is the general statement:

Proposition 1.17. *Suppose* $\gcd(n, 10) = 1$ *and write the prime factorization of* n *as*

$$n = p_1^{h_1} p_2^{h_2} \cdots p_r^{h_r}.$$

If ℓ_i *denotes the period of* $1/p_i^{h_i}$, *then the period of* $1/n$ *is* $\mathrm{lcm}(\ell_1, \ell_2, \ldots, \ell_r)$.

This proposition encourages us to look more closely at the periods of the fractions $1/p^k$ (with $p \neq 2, 5$). Here's the final result of this section:

Theorem 1.18. *Let* $p \neq 2, 5$ *be prime and let* $k \geqslant 1$. *Let* ℓ *be the period of* $1/p$ *and write*

$$10^{p-1} - 1 = p^b s, \tag{1.7}$$

where s *is not a multiple of* p. *Then the period of* $1/p^k$ *is given by*

$$\begin{cases} \ell & \text{if } 1 \leqslant k \leqslant b \\ \ell p^{k-b} & \text{if } k > b \end{cases}.$$

Unlike the previous theorems, I won't attempt to discuss a proof of this one because it's more complicated the proofs of our previous results. Instead, I refer the interested reader to [Ros, Lemma 4] for a proof, which relies on the theory of modular arithmetic.

I will make a few of remarks about this theorem, however. The first is that one knows $b \geqslant 1$ by Euler's (or Fermat's) Theorem. The second remark is that one can show the highest power of p dividing $10^{p-1} - 1$ is actually the same as the highest power of p dividing $10^\ell - 1$ (the proof in the Appendix shows this), so one could alternatively replace (1.7) with $10^{\ell-1} = p^b s'$, where s' is not a multiple of p.

The third and most fascinating remark is that, if one looks at the data, it's actually *very* rare to find primes p where $b > 1$. Indeed, after $p = 3$ (for which (1.7) reads $10^2 - 1 = 3^2 \cdot 1$), the next prime for which $b > 1$ is $p = 487$, and the next one after that is $p = 56598313$. These primes are called *base 10 Weiferich primes*, and no one has ever discovered any others besides these three; in fact, it's an unsolved problem to determine whether their number is finite or infinite. That is, we don't know the answer to the following question: *Are there infinitely many primes* p *for which the periods of the decimal expansions of* $1/p$ *and* $1/p^2$ *are the same?*

Turning back to $1/n$ Question 1.5, we've now come to the very heart of the problem. Thanks to elegant multiplicative structures of the integers, namely prime factorization and Euler's Theorem, the period of $1/n$ is determined by the periods of the numbers $1/p^k$ and, by some further elegance in elementary number theory (see the Appendix), these are in turn determined by the periods of the numbers $1/p$. But if we're staring at some prime p, we know from the graph in Figure 1 that it's going to be difficult to predict the period of $1/p$.

p	**Decimal expansion of $1/p$**
7	$0.\overline{142857}$
17	$0.\overline{0588235294117647}$
19	$0.\overline{052631578947368421}$
23	$0.\overline{0434782608695652173913}$
29	$0.\overline{0344827586206896551724137931}$
47	$0.\overline{0212765957446808510638297872340425531914893617}$
59	$0.\overline{0169491525423728813559322033898305084745762711864406779661}$
61	$0.\overline{016393442622950819672131147540983606557377049180327868852459}$

Table 3: Some decimal expansions of $1/p$ with period $p-1$

It's important to emphasize the role played by the prime factorization of n here. Factoring large integers is notoriously difficult and so starting with, say, $n = 215071$ and arriving at the prime factorization $n = 449 \cdot 479$ is already a challenge. But there's an added layer of difficulty on top of that, because we still need to pin down the periods of $1/449$ and $1/479$, and that's probably not going to be easy.[5] This is what I find so fascinating about this whole quest: after all of the careful, structured analysis we've done, at the heart of our problem resides the chaotic realm depicted in Figure 1, where the secrets are far from being uncovered.

2 What can we learn from the digits of $1/p$?

In this part, we'll turn our focus to the decimal expansions of $1/p$, for a prime $p \neq 2, 5$, which we've just identified as the central figures in determining the periods of the expansions of $1/n$. Instead of starting with a prime p and asking what the period of $1/p$ will be, we're going to suppose that we're looking at a value of p for which the period is as long as possible, namely $p-1$. The (admittedly vague) question will then be:

Question 2.1. *If $p \neq 2, 5$ is a prime and the decimal expansion of $1/p$ has period $p-1$ (or equivalently, if 10 is a primitive root of p), what can be said about the repeating sequence of $p-1$ digits?*

Thus throughout this part, p will always be a prime that has 10 as a primitive root. When $p = 4k+3$ (for some integer k), we'll see that the digits of the expansion of $1/p$ are linked to profound ideas in number theory. Unlike Part 1, my goal in Part 2 won't be to prove anything but simply these convey these sophisticated ideas to the uninitiated reader.

2.1 How random-looking are the digits of $1/p$?

Let's start by looking at some data again. Table 3 lists the first few values of p for which $1/p$ has period $p-1$. Obviously this string of $p-1$ digits is going to very large when p is large, and it's a lot to take in with the naked eye. For instance, let's look at the case of $p = 47$:

$$1/47 = 0.\overline{0212765957446808510638297872340425531914893617}.$$

The first few digits are going to be fairly predictable, because we know $1/47 \approx 1/50 = 0.02$. But past a certain point the digits start looking, well, kind of *random*.

[5] I wanted to avoid discussing the use of *congruences* in this article, but if you know about them, they can make the process slightly easier. For instance, since the period of $1/479$ divides $478 = 2 \cdot 239$, you know it's either $1, 2, 239$, or 478. So by Proposition 1.6, you can check whether 479 divides each of $10-1$, 10^2-1, and $10^{239}-1$ (there's no need to check $10^{478}-1$; see Theorem 1.14), which means determining whether any of $10, 10^2$, or 10^{239} are congruent to 1 modulo 479. There are other number-theoretic tricks that can help as well, but in general these computations are still tedious.

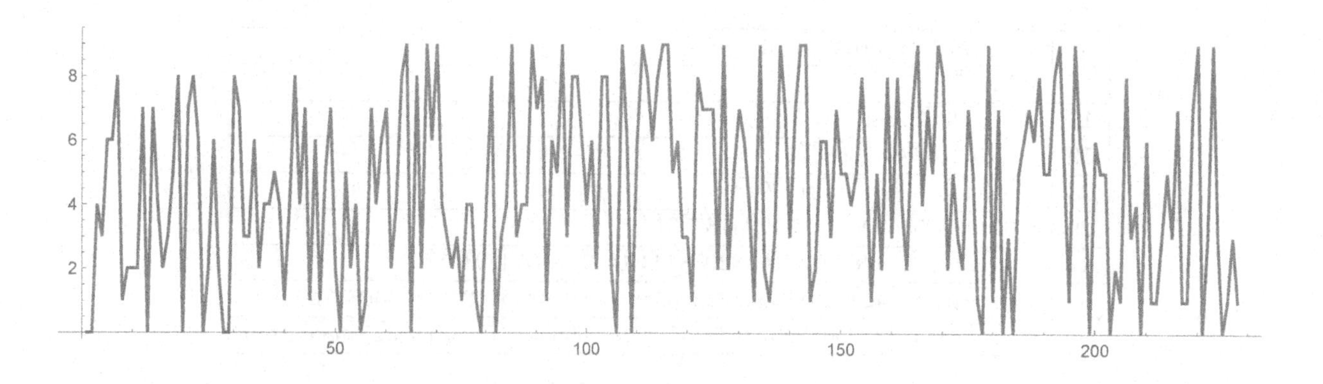

Figure 2: Plot of the 228-digit repeating sequence in the expansion of 1/229

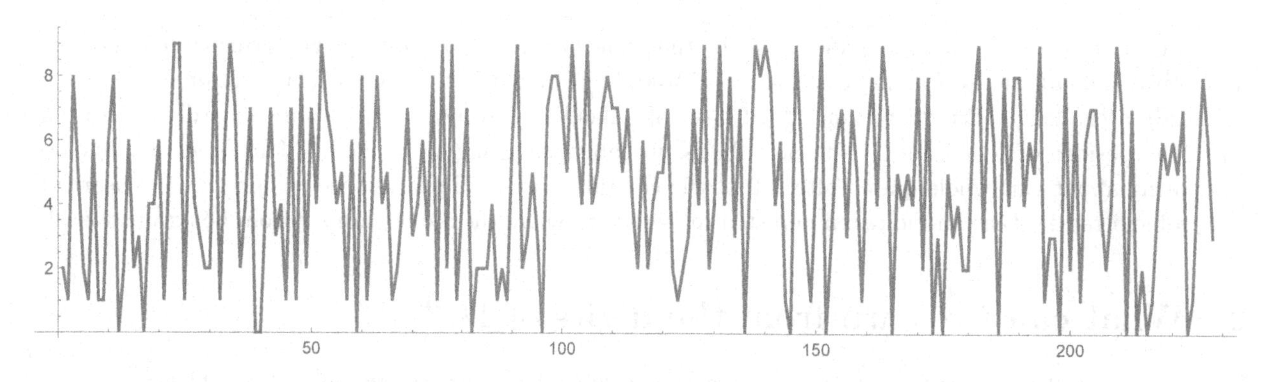

Figure 3: Plot of a random sequence of 228 decimal digits

Let me clarify two things about what I just said. First of all, the digits of 1/47 (and $1/p$ in general) are definitely *not* random. Whatever procedure I follow to calculate the decimal expansion of 1/47, as long as it's correct I'm going to get the same repeating sequence of 46 digits every time. I'm just saying that past a certain point the digits "look" random to me, and that's the second thing to clarify: this is just my opinion. You might look at those digits and feel differently.

If you don't feel this way, let me try to sway you to my side a bit. In the five rows below are strings of 35 digits. One of the rows is a string of 35 digits taken from the middle of the expansion of 1/131 (which has period 130), one of them is from the middle of the expansion of 1/229 (which has period 228), and the other three rows are strings of 35 random digits. *Which of these three rows are random strings?*

41984732824427480916030534351145038

18782860059580793714132782255045931

56165863608677584594706576115269330

64017277591261236900963254209039020

69868995633187772925764192139737991

It's pretty tough to tell.[6]

For a more visual demonstration, we can plot the digits in the sequence of repeating digits in the expansion of $1/p$, and compare the plot with a random sequence of digits. The 228-digit sequence coming from the expansion of 1/229 is plotted in Figure 2, and a sequence of 228 randomly-generated digits is given in Figure

[6] The answer is that the first row comes from 1/131 and the last row comes from 1/229.

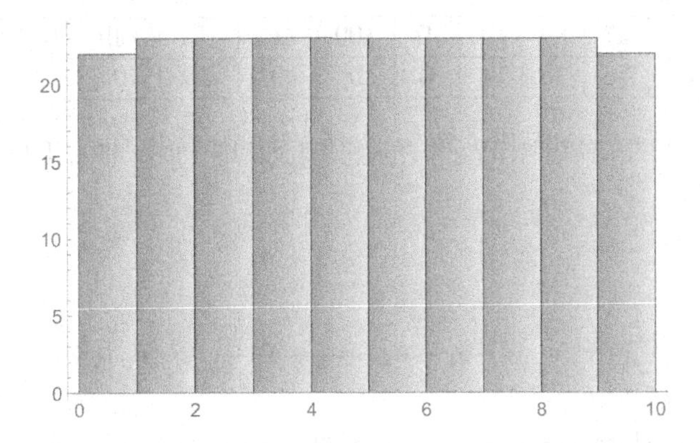

Figure 4: Digit frequency in the 228-digit repeating sequence in the expansion of $1/229$

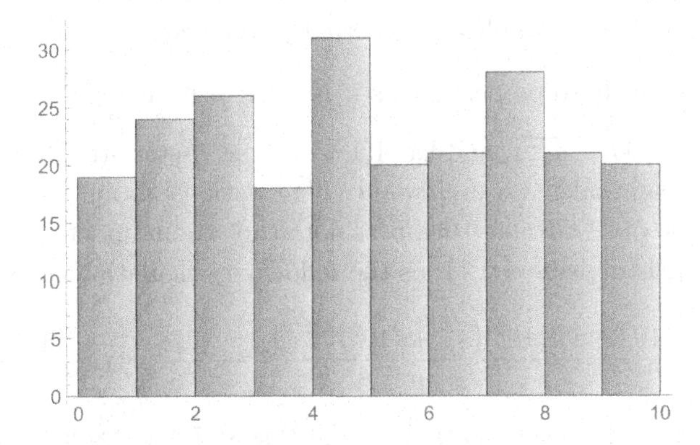

Figure 5: Digit frequency in a random sequence of 228 decimal digits

3.[7] We can see in Figure 2 that the first couple of digits in the sequence are 0's, but after that I feel like the two plots have more or less the same character.

Now this might be starting to feel like a setup to you: I've spent these last couple paragraphs trying to convince you that the digits of $1/p$ look like a random sequence, and now you're just waiting for the moment where I reveal a hidden bias in the sequence that makes it seem a lot less random. If this is your suspicion, well, you're right, that's what I'm about to do... And in fact there are several ways I can do this. For instance, I can take the sequences in Figures 2 and 3 and make a histogram of how many times each digit appears. This is done in Figures 4 and 5, and one can see there that the digits in the expansion of $1/229$ are "suspiciously" well-distributed when compared with the random sequence. Alternatively, I can cite what is known as *Midy's Theorem* (see [Ros], for instance) which implies that if $1/p = 0.\overline{\mathtt{d}_1\mathtt{d}_2\ldots\mathtt{d}_{p-1}}$, then

$$\mathtt{d}_i + \mathtt{d}_{\frac{p-1}{2}+i} = 9. \tag{2.1}$$

This says that the ith digit and the $(\frac{p-1}{2}+i)$th digit are inextricably linked; in particular, the first half of the sequence completely determines the second half of the sequence.

But the hidden bias I want to tell you about is one that I feel is much deeper than the ones mentioned above, and really gives the impression that the digits of $1/p$ harbor some very startling secrets. This bias was first demonstrated by the number theorist Kurt Girstmair, and we'll describe his result in terms of another simple statistical test: if $\mathtt{e}_1, \mathtt{e}_2, \ldots \mathtt{e}_{p-1}$ is *any* string of $p-1$ digits, we'll say that the *A-statistic* of this

[7] To be more precise, the latter sequence is a computer-generated *pseudorandom* sequence, which is meant to mimic a random sequence of digits in which each digit from 0 through 9 is chosen with uniform probability and chosen independently.

p	7	17	19	23	47	59	61	97	109	113	131	149	167	179	181	193	223
$A(1/p)$	11	0	11	33	55	33	0	0	0	0	55	0	121	55	0	0	77

Table 4: A-statistic applied to the repeating sequence in the expansion of $1/p$

string is the alternating sum

$$A := \sum_{j=1}^{p-1}(-1)^j \mathsf{e}_j = -\mathsf{e}_1 + \mathsf{e}_2 - \mathsf{e}_3 + \mathsf{e}_4 - \ldots - \mathsf{e}_{p-2} + \mathsf{e}_{p-1}.$$

What should we expect about the value of A when this string $\mathsf{e}_1, \mathsf{e}_2, \ldots \mathsf{e}_{p-1}$ is truly random? If p is large and the string is of digits is random then:

(a) The value of A is likely to be close to zero (since $p-1$ is even and the expected value of each digit is the same, namely 4.5), and yet it's unlikely to actually equal zero.

(b) The value of A is just as likely to be positive as it is to be negative.

Now suppose that $1/p = 0.\overline{\mathsf{d}_1\mathsf{d}_2\ldots\mathsf{d}_{p-1}}$, and let $A(1/p)$ be the A-statistic of the sequence $\mathsf{d}_1, \mathsf{d}_2, \ldots \mathsf{d}_{p-1}$. With a little bit of work, we can make two preliminary observations about $A(1/p)$. First of all, we should expect $A(1/p)$ to be a multiple of 11. To see this, first note that the number $(10^{p-1}-1)/p$, whose decimal representation is $\mathsf{d}_1\mathsf{d}_2\ldots\mathsf{d}_{p-1}$, is divisible by 11, as the following calculation shows:

$$\frac{10^{p-1}-1}{p} = \frac{(10^2)^{\frac{p-1}{2}}-1}{p} = \frac{(10^2-1)((10^2)^{\frac{p-3}{2}}+(10^2)^{\frac{p-5}{2}}+\cdots+1)}{p} = 11\cdot\frac{9(100^{\frac{p-3}{2}}+100^{\frac{p-5}{2}}+\cdots+1)}{p}.$$

Here we're using the fact that $p-1$ is even (so that $(p-1)/2$ is an integer) and that $p \neq 11$ (so the factor of 11 on the right side doesn't get cancelled out when we divide by p). On the other hand, a well-known test for divisibility by 11 says that $(10^{p-1}-1)/p$ is a multiple of 11 if and only if the A-statistic $A(1/p) = \sum_{i=1}^{p-1}(-1)^i \mathsf{d}_i$ is a multiple of 11.

The other observation we can make is that if the $p-1$ is a multiple of 4, then $A(1/p)$ will be 0; this is less obvious, but it follows from Midy's Theorem above. Indeed, in this case $(p-1)/2$ will be even, which implies that i and $(p-1)/2+i$ are either both even or both odd. So when we split the alternating sum $A(1/p)$ into positive terms and negative terms, the formula (2.1) implies that

$$A(1/p) = \sum_{\substack{1\leqslant i\leqslant p-1 \\ i \text{ even}}} \mathsf{d}_i - \sum_{\substack{1\leqslant i\leqslant p-1 \\ i \text{ odd}}} \mathsf{d}_i = \frac{p-1}{4}\cdot 9 - \frac{p-1}{4}\cdot 9 = 0.$$

Notice that both of these observations are contrary to the criterion (a) that we have above for "randomness" of the sequence $\mathsf{d}_1, \mathsf{d}_2, \ldots \mathsf{d}_{p-1}$.

Now let's look at some actual data. Table 4 lists, for various p, the values of the A-statistic $A(1/p)$. Our two observations are visible in this table, but notice that there is more than we anticipated: in this small data set we see that, when p has the form $4k+3$, the statistic $A(1/p)$ is always a *positive* multiple of 11, contrary to our criterion (b) for randomness. In fact this statement holds in general, and this follows from Girstmair's theorem:

Theorem 2.2 (Girstmair 1994). *For a prime number p, suppose that 10 is a primitive root of p and let $1/p = 0.\overline{\mathsf{d}_1\mathsf{d}_2\ldots\mathsf{d}_{p-1}}$. The A-statistic of the sequence $\mathsf{d}_1, \mathsf{d}_2, \ldots \mathsf{d}_{p-1}$ is given by*

$$A(1/p) = \begin{cases} 0 & \text{if } p = 4k+1 \text{ for some } k \\ 11N_p & \text{if } p = 4k+3 \text{ for some } k, \end{cases}$$

where N_p is a certain positive integer depending upon p.

This might seem like only a slight improvement upon our two observations made above, and not a very interesting conspiracy. But the statement in Theorem 2.2 is only the tip of the iceberg. As we discuss in the next section, Girstmair showed is that the integer N_p is one that has interested number theorists for centuries.

2.2 Binary quadratic forms

To get an idea about the significance of the numbers N_p appearing in Theorem 2.2, we need to consider some questions that seem far removed from our earlier ones. These are questions that have intrigued number theorists since at least the time of Fermat, and perhaps the simplest of them is the following: Which integers $n \geqslant 0$ are a sum of two squares? That is, for which n do there exists integers x, y such that $n = x^2 + y^2$? Fermat proved a theorem that gave a criterion for answering this question completely and, perhaps unexpectedly, the key step in his proof is to understand which *primes* can be written as a sum of two squares.[8] Similarly, Fermat was able to determine when a prime could be written in the form $x^2 + 2y^2$, and from this he could answer the more general question of which positive integers had this form. He could also tell when a prime, and hence any integer, had the form $x^2 + 3y^2$. But the next[9] logical question was to ask which primes could be written in the form $x^2 + 5y^2$, and Fermat was unable to provide an answer.

Successive generations of mathematicians, especially Euler, Legendre, Lagrange, and Gauss, took up the task of (verifying and) continuing Fermat's work, and were drawn into a larger landscape of equations, one that turned out to be surprisingly difficult to navigate. This larger landscape is the theory of *integral binary quadratic forms*, which are expressions of the form

$$ax^2 + bxy + cy^2,$$

where a, b, c are some fixed integers and x, y are the indeterminates. We'll usually just call these "quadratic forms" or even just "forms." The question, say, of deciding which integers are a sum of two squares can be phrased of asking which integers n are "represented" by the quadratic form $x^2 + y^2$. More generally, we say n is represented by the quadratic form $ax^2 + bxy + cy^2$ if there exist integers x, y such that

$$ax^2 + bxy + cy^2 = n.$$

Motivated by Fermat's work, one is lead to ask:

Question 2.3. *Which prime numbers are represented by a given quadratic form $ax^2 + bxy + cy^2$?*

In answering this question, it's worth noting that certain quadratic forms are less deserving of our attention than others. Specifically, if the coefficients a, b, c are all share a nontrivial factor, then any integer represented by that form must also be divisible by that factor. As one example, the only prime that the form $5x^2 - 15xy + 25y^2$ could (possibly) represent is 5 because the coefficients (and hence all of the integers it represents) are divisible by 5. As second example, the form $4x^2 - 8xy + 12y^2$ can never represent any prime because 4 divides all of the coefficients. For this reason, we'll restrict our attention to so-called *primitive* quadratic forms, which are those satisfying the condition $\gcd(a, b, c) = 1$.

Now certain quadratic forms can be seen to represent exactly the same integers, for the simple reason that they are related by a reversible substitution. To demonstrate this, let's show that the simple form $q_1(x, y) = x^2 + y^2$ represents exactly the same integers as the more complicated form

$$q_2(X, Y) = 13X^2 - 16XY + 5Y^2.$$

[8] The answer, it turns out, is that the prime must be either 2 or have the form $4k + 1$; but we're going to focus more on the process of obtaining the answer, rather than the actual answer. For the answer to which integers (prime or composite) are a sum of two squares, an internet search or a book on elementary number theory can get you started.

[9] The question of which primes have the form $x^2 + 4y^2 = x^2 + (2y)^2$ has the nearly the same answer as primes of the form $x^2 + y^2$ (just omit the prime 2), so this question isn't so interesting.

First notice that if we let $x = 2X - Y, y = -3X + 2Y$ and plug in this into the form $q_1(x, y)$, we find

$$x^2 + y^2 = (2X - Y)^2 + (-3X + 2Y)^2 = 13X^2 - 16XY + 5Y^2.$$

Thus the fact that, say, 61 is represented by $q_2(X, Y)$ will imply it's also represented by $q_1(x, y)$: starting from $61 = q_2(7, 8)$, we let $x_0 = 2\cdot7-8 = 6$, $Y_0 = -3\cdot7+2\cdot8 = -5$, and conclude (as you can verify directly) that $61 = q_1(6, -5)$. Conversely, since

$$\begin{cases} x = 2X - Y \\ y = -3X + 2Y \end{cases} \implies \begin{cases} X = 2x + y \\ Y = 3x + 2y \end{cases}$$

we can go in the opposite direction:

$$13X^2 - 16XY + 5Y^2 = 13(2x + y)^2 - 16(2x + y)(3x + 2y) + 5(3x + 2y)^2 = x^2 + y^2$$

Given the representation $29 = 5^2 + 2^2$, we can let $X_0 = 2 \cdot 5 + 2 = 12$, $Y_0 = 3 \cdot 5 + 2 \cdot 2 = 19$, and obtain the representation $29 = q_2(12, 19)$.

In general, we can make the substitution $x = \alpha X + \beta Y$, $y = \gamma X + \delta Y$ in a quadratic form $q(x, y)$ to obtain a new quadratic form $Q(X, Y)$. If $\alpha, \beta, \gamma, \delta$ are all integers, then x, y will be integers whenever X, Y are integers; thus any n represented by $Q(X, Y)$ is represented by $q(x, y)$. In order to go in the opposite direction, we need to say that when x, y are integers, so are X, Y; this will be true when the integers $\alpha, \beta, \gamma, \delta$ satisfy the condition $\alpha\delta - \beta\gamma = 1$. [10] This motivates the following:

Definition 2.4. *Two integral binary quadratic forms $q(x, y)$ and $Q(X, Y)$ are called* equivalent *if we can transform $q(x, y)$ into $Q(X, Y)$ by using a substitution of the form $x = \alpha X + \beta Y$, $y = \gamma X + \delta Y$, where $\alpha, \beta, \gamma, \delta$ are integers that satisfy $\alpha\delta - \beta\gamma = 1$.*

Given a form $q(x, y)$, the collection of all forms that are equivalent to $q(x, y)$ is called its equivalence class.

Now suppose that we're handed two quadratic forms $q(x, y)$ and $Q(X, Y)$; is there a way to tell whether they're equivalent? We could try using the definition directly, which would mean searching for an appropriate substitution that will transform one into the other. But there are actually infinitely many possible substitutions, and this kind of guess-and-check approach doesn't present a clear way forward. Instead, mathematicians seized upon an easier feature of each quadratic form, its *discriminant.* The discriminant of $q(x, y) = ax^2 + bxy + cy^2$ is the integer $D = b^2 - 4ac$, and the story is particularly nice when $D < 0$. So we will only focus upon forms with negative discriminant; we will also restrict ourselves to forms where $a > 0$. A form $q(x, y)$ with $D < 0$ is called *positive definite*, because $q(x, y) > 0$ whenever $(x, y) \neq (0, 0)$.

The importance of the discriminant is that, as one can show by a tedious but straightforward check, equivalent quadratic forms always have equal discriminants. (For instance, you can verify in the example above that both $q_1(x, y)$ and $q_2(X, Y)$ both have discriminant -4.) To see how useful this is, notice the two quadratic forms

$$q(x, y) = 7x^2 - 3xy + y^2, \qquad Q(X, Y) = 2X^2 + 2XY + Y^2$$

are not equivalent, simply because the discriminant of $q(x, y)$ is -19 while the discriminant of $Q(X, Y)$ is -4. In other words, no matter how many substitutions as in Definition 2.4 we try, we'll never be able to turn $q(x, y)$ into $Q(X, Y)$!

In this way, the discriminant is a very handy tool for showing that certain pairs of quadratic forms are not equivalent. It naturally leads us to wonder whether the discriminant can help us completely solve the problem of equivalence:

---

[10] More generally, this will work exactly when $\alpha\delta - \beta\gamma = \pm1$. But it was Gauss who suggested using substitutions satisfying the stricter condition $\alpha\delta - \beta\gamma = 1$, and that the relations between various quadratic forms are illuminated much better if one makes this small change. In any case, this will be the most relevant type of substitution for our purposes.

Question 2.5. *If two quadratic forms $q(x,y)$ and $Q(X,Y)$ both have the same discriminant D, are they necessarily equivalent?*

Let's pause at this point to acknowledge that, with all this terminology about quadratic forms and their classification, it might feel like we've diverged from the original spirit of Fermat's questions. But it turns out that the preceding question has a direct impact upon Fermat's interests: if the answer to Question 2.5 is yes, it is *much* easier to answer Question 2.3 regarding which primes—and which integers in general—are represented by a quadratic form of discriminant D.

Let me give a vague idea of what I mean when I say that Question 2.3 is "easier" to answer in this case. If we have a quadratic form $q(x,y)$ of discriminant D and a prime p, there is a certain *necessary* condition, let's call it $\mathcal{C}_D$ for short, that must be satisfied by p in order to say $q(x,y)$ can represent p. While I can't state it precisely here, a few examples can give a hint about the general character of $\mathcal{C}_D$:

- $\mathcal{C}_{-4}$: If $q(x,y)$ has $D=-4$ and $p \neq 2$, p must have the form $4k+1$ (for some integer k) in order to be represented by $q(x,y)$.
- $\mathcal{C}_{-8}$: If $q(x,y)$ has $D=-8$ and $p \neq 2$, p must have the form $8k+1$ or $8k+3$ in order to be represented by $q(x,y)$.
- $\mathcal{C}_{-12}$: If $q(x,y)$ has $D=-12$ and $p \neq 2,3$, p must have the form $3k+1$ in order to be represented by $q(x,y)$.
- $\mathcal{C}_{-20}$: If $q(x,y)$ has $D=-20$ and $p \neq 2,5$, p must have the form $20k+1, 20k+3, 20k+7$, or $20k+9$ in order to be represented by $q(x,y)$.

For a given discriminant D, $\mathcal{C}_D$ is not too difficult to write down if one knows the so-called Law of Quadratic Reciprocity, a standard topic in undergraduate number theory. The upshot is that when the answer to Question 2.5 is yes, the condition $\mathcal{C}_D$ is also a *sufficient* that guarantees $q(x,y)$ will represent p; when the answer is no, $\mathcal{C}_D$ is not sufficient and more complicated ideas are required to determine whether $q(x,y)$ will represents p.

For instance, it turns out that all primitive quadratic forms of discriminant -4 are equivalent, and the same is true of quadratic forms of discriminant -8 and -12. Thus the conditions $\mathcal{C}_{-4}, \mathcal{C}_{-8}$, and $\mathcal{C}_{-12}$ tell us which primes can be written in the form x^2+y^2, x^2+2y^2, or x^2+3y^2. However, not all quadratic forms of discriminant -20 turn out to be equivalent. For instance, one can show that x^2+5y^2 and $2x^2+2xy+3y^2$ are two inequivalent forms of discriminant -20; furthermore (ignoring the primes 2 and 5), the form x^2+5y^2 only represents the primes of the form $20k+1$ and $20k+9$ while $2x^2+2xy+3y^2$ only represents the primes of the form $20k+3$ and $20k+7$. Hence $\mathcal{C}_{-20}$ is not a sufficient condition for telling when a form of discriminant -20 represents a prime. In light of this fact, it seems less surprising that Fermat had difficulty determining the primes of the form x^2+5y^2.

Getting back to the basic concern of Question 2.5, it might help to view the landscape of all (primitive positive definite) quadratic forms as being divided up into infinitely many separate "states." Each state consists of all of the forms having the same discriminant D, and any given quadratic form $q(x,y)$ belongs to exactly one of these states. In fact, since equivalent forms share the same discriminant, the entire equivalence class of $q(x,y)$ is contained within a single state. If we think of each equivalence class as a "county," then each state is divided up into counties. In this map-making analogy, Question 2.5 asks whether every state contains only one county. In the states with $D=-4,-8,-12$, the answer is yes but in the state with $D=-20$, there turn out to be two counties. This means the general problem of determining whether two forms are equivalent—and thus which primes are represented by a form—is difficult.

One bit of good news, though, is that there happen to be only finitely many equivalence classes corresponding to every discriminant, i.e., each state has only finitely many counties. This is not an easy fact to prove, but once we have it we can make the following definition:

Definition 2.6. *For an integer $D < 0$, the* class number $h(D)$ *is the number of distinct equivalence classes of quadratic forms having discriminant D.*

Regarding the examples above, we have $h(-4) = h(-8) = h(-12) = 1$ and $h(-20) = 2$.

We are now *finally* ready to state Girstmair's Theorem 2.2 in a more precise manner. Remember that theorem? It had something to do with the decimal expansion of $1/p$... that seems a world away from us now, after all of this talk about quadratic forms. But that perceived distance is exactly what makes his result so surprising:

Theorem 2.7 (Girstmair 1994). *For a prime number p, suppose that 10 is a primitive root of p and let $1/p = 0.\overline{\mathtt{d}_1\mathtt{d}_2\ldots\mathtt{d}_{p-1}}$. The A-statistic of the sequence $\mathtt{d}_1, \mathtt{d}_2, \ldots \mathtt{d}_{p-1}$ is given by*

$$A(1/p) = \begin{cases} 0 & \text{if } p = 4k+1 \text{ for some } k \\ 11h(-p) & \text{if } p = 4k+3 \text{ for some } k, \end{cases}$$

where $h(-p)$ is the class number of all primitive binary quadratic forms having discriminant $-p$.

For instance, we have $A(1/7) = -1 + 4 - 2 + 8 - 5 + 7 = 11$. Therefore $h(-7) = 1$, meaning that all primitive quadratic forms of discriminant -7 are equivalent to each other. We can make this statement more concrete by singling one out: all forms with $D = -7$ are equivalent to $x^2 + xy + 2y^2$. So this means it's "easy" to determine which primes are represented forms with $D = -7$. For instance, since $1^2 + 1 \cdot 2 \cdot 2^2 = 7$, the form $x^2 + xy + 2y^2$ represents 7, and hence so do all forms with $D = -7$. Moreover, the condition $\mathcal{C}_{-7}$ reads:

- $\mathcal{C}_{-7}$: If $q(x,y)$ has $D = -7$ and $p \neq 7$, p must have the form $7k+1$, $7k+2$, or $7k+4$ in order to be represented by $q(x,y)$.

Since $h(-7) = 1$, the condition $\mathcal{C}_{-7}$ is sufficient, meaning that a quadratic form with $D = -7$ will represent the primes $2, 7, 11, 23, 29, 37, 43, \ldots$ and will not represent the primes $3, 5, 13, 17, 19, 31, 41, \ldots$.

Since we have $A(1/19) = 11$ and thus $h(-19) = 1$, a similar analysis can tell us which primes are represented by forms with $D = -19$. On the other hand, we see that $A(1/23) = 33$, so that $h(-23) = 3$. Therefore it takes more work, and more techniques than someone like Fermat would have had available, to determine which primes are represented by a form such as $x^2 + xy + 6y^2$, which has $D = -23$.

Now I don't want to give the impression that both of these facts are new, because the values of $h(-7)$, $h(-19)$, and $h(-23)$ have been known for a long time. But they were computed using methods that required significantly more background to carry out. What Theorem 2.7 shows is that, during all those years, there were some lowly decimal expansions who also knew these values, and they held their tongue about it all the way until the 1990s.

3 Further directions

In this conclusion, I'll indicate a few things that I didn't include in the previous parts, as well as some places where an interested reader can find out more.

3.1 Beyond base 10

One of the goals of this article is to bring out some ideas from number theory in a context that is more familiar and less intimidating than the general setting. Specifically, that context is our usual way of representing numbers using the decimal system, i.e., using the base 10. But we can also consider the *binary* expansion of $1/n$ (corresponding to base 2), or the *hexidemical* expansion of $1/n$ (corresponding to base 16), or more generally the base b expansion of $1/n$ for any integer $b \geqslant 2$.

The facts we proved in Part 1 continue hold if we consider the base b expansion of $1/n$:

- The base b expansion of $1/n$ is finite if and only if b and n are divisible by exactly the same prime numbers.
- If $\gcd(n, b) = 1$, then the base b expansion of $1/n$ is infinite and (purely) periodic.
- The length ℓ of the shortest repeating sequence in the base b expansion of $1/n$ is the smallest integer $\ell \geqslant 1$ such that $b^\ell - 1$ is divisible by n.
- When p is prime, this length ℓ divides $p-1$, and we have $\ell = p - 1$ exactly when b is a primitive root of p.

These facts are proved with the help of the full version of Euler's Theorem, which states that n divides $b^{\phi(n)} - 1$ whenever $\gcd(n, b) = 1$. For more about these base-b expansions, see [HW].

There is also a version of Conjecture 1.12 for general b, called Artin's Primtive Root Conjecture; in particular, it predicts that, when b is not a square, there should exist infinitely many primes p having b as a primitive root. See [Mur] for more on this conjecture.

Finally, Girstmair actually stated his results not for decimal expansions, but in terms of a general base-b expansion. His result states that if $p = 4k + 3$ and b is a primitive root of p, and hence the base-b expansion of $1/p$ has period $p-1$, then the A-statistic of the repeating digits in this expansion equals $(b+1)h(-p)$. This allows us to compute other class numbers that can't be computed with decimal expansions. For instance, we can't use the A-statistic of the decimal expansion of $1/11$ to compute $h(-11)$, but we could use the binary expansion of $1/11$ to do so because 2 is a primitive root of 11.

Girstmair's most accessible account of his result is [Gir2]. The result there is a special case of more technical work in [Gir1]. The later works [Gir3] and [MT] give some extensions, such as the fact that one can compute $h(-p)$ using the base-b expansion of $1/p$ when $p = 4k + 3$ and the period is $(p-1)/2$. (One doesn't use the A-statistic in that case, but something similarly simple.)

3.2 The virtues of abstract algebra

Most of the results in Part 1 can be proved more easily if one knows some abstract algebra, namely at the level of the rings $\mathbb{Z}/n\mathbb{Z}$ and their groups of units $(\mathbb{Z}/n\mathbb{Z})^\times$. The specific link that one needs is Proposition 1.6, which connects the period of $1/n$ (when $\gcd(n, 10) = 1$) to the smallest exponent ℓ such that n divides $10^\ell - 1$. In the language of abstract algebra, this ℓ is the same as the order of the element $[10]$ in the multiplicative group $(\mathbb{Z}/n\mathbb{Z})^\times$. So much of our concern in Part 1 is really about trying to pin down this order.

If one understands this translation and is fluent in the language, things become easier. For instance, Euler's Theorem follows very quickly from a general theorem of Lagrange when applied to the finite group $(\mathbb{Z}/n\mathbb{Z})^\times$, whose order is $\phi(n)$. Moreover, the so-called *Chinese Remainder Theorem* says that if the prime factorization of n is

$$n = p_1^{h_1} p_2^{h_2} \cdots p_r^{h_r},$$

then we have a ring isomorphism

$$\mathbb{Z}/n\mathbb{Z} \simeq (\mathbb{Z}/p_1^{h_1}\mathbb{Z}) \oplus (\mathbb{Z}/p_2^{h_2}\mathbb{Z}) \oplus \cdots \oplus (\mathbb{Z}/p_r^{h_r}\mathbb{Z}),$$

which one can restrict to the units to give the following isomorphism of groups:

$$(\mathbb{Z}/n\mathbb{Z})^\times \simeq (\mathbb{Z}/p_1^{h_1}\mathbb{Z})^\times \times (\mathbb{Z}/p_2^{h_2}\mathbb{Z})^\times \times \cdots \times (\mathbb{Z}/p_r^{h_r}\mathbb{Z})^\times.$$

With this isomorphism in hand, one is not too far away from being able to deduce Proposition 1.17.

In this way, abstract algebra handles the elegant structural reductions that we made in Part 1 with great ease. But it doesn't have as much to offer in terms of understanding the chaotic graph in Figure 1. In other words, whether we call it the period of the decimal expansion of $1/p$ or the order of $[10]$ in $(\mathbb{Z}/p\mathbb{Z})^\times$, we still have trouble understanding how this varies as p gets larger!

3.3 Another interpretation of $h(-p)$

In addition to counting equivalence classes of binary quadratic forms of discriminant $D = -p$, there is another interpretation of the class number $h(-p)$ that arises in Girstmair's theorem: it is measures the size of the so-called *ideal class group* of the imaginary quadratic field $\mathbb{Q}(\sqrt{-p})$.

Let me try to convey what this means. For a prime p, we can consider complex numbers of the form $a + b\sqrt{-p}$, where a and b are both *rational* numbers; note that this can also be written as as $a + bi\sqrt{p}$, where $i = \sqrt{-1}$. This is a subset of the set of all complex numbers that is typically denoted as $\mathbb{Q}(\sqrt{-p})$. With some effort, one can verify that if we take two numbers in $\mathbb{Q}(\sqrt{-p})$, we can add, subtract, or multiply them and we'll get another number in $\mathbb{Q}(\sqrt{-p})$; moreover, if one of them is not zero, we can divide it into the other one and again obtain something in $\mathbb{Q}(\sqrt{-p})$. We say that $\mathbb{Q}(\sqrt{-p})$ is closed under addition, subtraction, multiplication, and division, and these nice properties make $\mathbb{Q}(\sqrt{-p})$ into a *field* in the parlance of abstract algebra. More generally, we can replace p by any integer d which is squarefree (i.e., is a product of distinct primes), and we'll obtain a field denoted as $\mathbb{Q}(\sqrt{-d})$. Such fields are called *imaginary quadratic fields.*

Although they seem fairly abstract, these fields receive a lot of attention in number theory because they are related to statements about integers. For instance, if n is a positive, can it be written in the form $x^2 + 7y^2$ for some integers x, y? This is equivalent to asking if the equation

$$n = x^2 + 7y^2 \tag{3.1}$$

has integer solutions x, y. But an inspired factorization turns this equation into

$$n = (x + y\sqrt{-7})(x - y\sqrt{-7}). \tag{3.2}$$

Notice that $x \pm y\sqrt{-7}$ belong to the field $\mathbb{Q}(\sqrt{-7})$, which means we've turned our original question about *adding* certain kinds of integers into a question about *multiplying* certain kinds of numbers in $\mathbb{Q}(\sqrt{-7})$.

In fact, in (3.2) we're not just multiplying any kind of numbers in $\mathbb{Q}(\sqrt{-7})$; the assumption is that x, y are integers, and not just rational. This implies that the numbers $x \pm y\sqrt{-7}$ are what are called *algebraic integers.* I won't go into the general definition of algebraic integers, but if we let R_7 denote the collection of all algebraic integers inside the field $\mathbb{Q}(\sqrt{-7})$, then I will say the following: R_7 is closed under addition, subtraction, and multiplication, but *not* division. For this reason, R_7 is called a *ring.*

In general, one can consider the subset of all algebraic integers R_d inside $\mathbb{Q}(\sqrt{-d})$, and R_d will always be a ring. Now the world's most famous ring is $\mathbb{Z}$, the ring of integers, and one can ask how much a ring of "algebraic integers" like R_d resembles the ring of "actual integers" $\mathbb{Z}$. In particular, one incredibly useful feature of the ring $\mathbb{Z}$ is its unique factorization property: integers other than 0 and ± 1 can be factored uniquely into a product of primes (along with a factor of -1 if the integer is negative). One can also define a notion of "primes" in the rings R_d. So the question arises as to whether most elements of R_d—that is, elements besides 0, ± 1, and other numbers with complex absolute value 1—can be written as a product of prime elements in an (essentially) unique way. If so, we call R_d a "unique factorization domain", or simply a UFD.

So one motivation for wanting R_d to be a UFD is that it looks more like the usual ring of integers $\mathbb{Z}$ that we know and love. But there is another reason: it makes certain problems about integers easier to solve. A perfect example of this is equation (3.1) above. If we knew that R_7 were a UFD, we could turn (3.1) into (3.2) and use unique factorization in R_7 to determine for which n there are integer solutions x, y.

Sadly, it turns out that R_d is not always a UFD. But R. Dedekind found a way to partially repair the situation by introducing *ideals,* which are certain nice subsets of R_d. One can show that these ideals factor (in sense I won't define) uniquely into so-called *prime ideals,* which are ideals in R_d that have "prime-like" properties. Moreover, by investigating relations between these ideals, one can create from them a supplementary object known as the *ideal class group,* which is a finite set (in fact, a finite group). The size of this group is a positive integer called the *ideal class number* of the imaginary quadratic field $\mathbb{Q}(\sqrt{-d})$, and we denote it as $h(-d)$.

Complicated thought it may be, the great thing about the ideal class number of $\mathbb{Q}(\sqrt{-d})$ is that it tells us whether R_d is a UFD. More precisely, R_d is a UFD if and only if the ideal class number $h(-d) = 1$. In that case, we're able to put aside the more intricate notions of ideals and prime ideals, and instead just work with the actual elements and prime elements of R_d.

Now let's get to the connection with quadratic forms: when p is a prime of the form $4k + 3$, the ideal class number of $\mathbb{Q}(\sqrt{-p})$ can be shown to equal to the number of equivalence classes of integral binary quadratic forms of discriminant $D = -p$. This is why it's safe to denote both of these things by $h(-p)$. So Girstmair's result says that when 10 is a primitive root of $p = 4k + 3$, the decimal expansion of $1/p$ can be used to calculate the ideal class group of $\mathbb{Q}(\sqrt{-p})$, and thus to determine whether the ring R_p is a UFD. In particular, the decimal expansion of $1/7$ can be used to conclude that R_7 is a UFD.

Imaginary quadratic fields are among the first examples of what are called *algebraic number fields*, which are the focus of algebraic number theory. Any introductory book on algebraic number theory will contain this material; the book [PD] presupposes less background, while books such as [Mar] or [Sam] assume a familiarity with much more abstract algebra. Another sophisticated introduction is the book [FT], which also explains the connection between the class numbers in §2.2 and the ideal class numbers discussed here. This is also explained in the excellent book [Cox], which is generally at a higher level than the preceding books, but also contains much historical material on quadratic forms that is easier to follow.

References

[Cox] D. A. Cox. *Primes of the form $x^2 + ny^2$*. Pure and Applied Mathematics (Hoboken). John Wiley & Sons, Inc., Hoboken, NJ, second edition, 2013. Fermat, class field theory, and complex multiplication. `http://dx.doi.org/10.1002/9781118400722`

[Dic] L. E. Dickson. *History of the theory of numbers. Vol. I: Divisibility and primality.* Chelsea Publishing Co., New York, 1966.

[FT] A. Fröhlich and M. J. Taylor. *Algebraic number theory*, volume 27 of *Cambridge Studies in Advanced Mathematics.* Cambridge University Press, Cambridge, 1993.

[Gir1] K. Girstmair. The digits of $1/p$ in connection with class number factors. *Acta Arith.* **67** (1994), 381–386.

[Gir2] K. Girstmair. A "popular" class number formula. *Amer. Math. Monthly* **101** (1994), 997–1001. `http://dx.doi.org/10.2307/2975167`

[Gir3] K. Girstmair. Periodische Dezimalbrüche—was nicht jeder darüber weiß. In *Jahrbuch Überblicke Mathematik, 1995*, pages 163–179. Vieweg, Braunschweig, 1995.

[HW] G. H. Hardy and E. M. Wright. *An introduction to the theory of numbers.* Oxford University Press, Oxford, sixth edition, 2008. Revised by D. R. Heath-Brown and J. H. Silverman, With a foreword by Andrew Wiles.

[HPS] J. Hoffstein, J. Pipher, and J. H. Silverman. *An introduction to mathematical cryptography.* Undergraduate Texts in Mathematics. Springer, New York, second edition, 2014. `http://dx.doi.org/10.1007/978-1-4939-1711-2`

[JP] R. Jones and J. Pearce. A postmodern view of fractions and the reciprocals of Fermat primes. *Math. Mag.* **73** (2000), 83–97. `http://dx.doi.org/10.2307/2691078`

[Lea] W. G. Leavitt. Repeating decimals. *College Math. J.* **15** (1984), 299–308. `http://dx.doi.org/10.2307/2686394`

[Mar] D. A. Marcus. *Number fields.* Springer-Verlag, New York-Heidelberg, 1977. Universitext.

[Mur] M. R. Murty. Artin's conjecture for primitive roots. *Math. Intelligencer* **10** (1988), 59–67. `http://dx.doi.org/10.1007/BF03023749`

[MT] M. R. Murty and R. Thangadurai. The class number of $\mathbb{Q}(\sqrt{-p})$ and digits of $1/p$. *Proc. Amer. Math. Soc.* **139** (2011), 1277–1289. `http://dx.doi.org/10.1090/S0002-9939-2010-10560-9`

[PD] H. Pollard and H. G. Diamond. *The theory of algebraic numbers.* Dover Publications, Inc., Mineola, NY, third edition, 1998.

[Ros] K. A. Ross. Repeating decimals: a period piece. *Math. Mag.* **83** (2010), 33–45. `http://dx.doi.org/10.4169/002557010X479974`

[Sam] P. Samuel. *Algebraic theory of numbers.* Translated from the French by Allan J. Silberger. Houghton Mifflin Co., Boston, Mass., 1970.

[SF] M. Shrader-Frechette. Complementary rational numbers. *Math. Mag.* **51** (1978), 90–98.

[Sil] J. H. Silverman. *A friendly introduction to number theory.* Pearson, fourth edition, 2012.

California State University, Fullerton, Department of Mathematics, Fullerton, CA 92834
Email address: `clyons@fullerton.edu`

The history of Algorithmic complexity

Audrey A. Nasar[1]
Borough of Manhattan Community College at the City University of New York

ABSTRACT: This paper provides a historical account of the development of algorithmic complexity in a form that is suitable to instructors of mathematics at the high school or undergraduate level. The study of algorithmic complexity, despite being deeply rooted in mathematics, is usually restricted to the computer science curriculum. By providing a historical account of algorithmic complexity through a mathematical lens, this paper aims to equip mathematics educators with the necessary background and framework for incorporating the analysis of algorithmic complexity into mathematics courses as early on as algebra or pre-calculus.

Keywords: Algorithm, Complexity, Discrete Mathematics, Mathematics Education

[1] anasar2@gmail.com

I. Introduction

Computers have changed our world. The exploding development of computational technology has brought algorithms into the spotlight and as a result, the analysis of algorithms has been gaining a lot of focus. An algorithm is a precise, systematic method for solving a class of problems. Algorithmic thinking, which is a form of mathematical thinking, refers to the thought processes associated with creating and analyzing algorithms. Both algorithms and algorithmic thinking are very powerful tools for problem solving and are considered to be key competences of students from primary to higher education (Cápay and Magdin, 2013). An integral component of algorithmic thinking is the study of algorithmic complexity, which addresses the amount of resources necessary to execute an algorithm. Through analyzing the complexity of different algorithms, one can compare their efficiencies, their mathematical characteristics, and the speed at which they can be performed.

The analysis of algorithmic complexity emerged as a scientific subject during the 1960's and has been quickly established as one of the most active fields of study. Today, algorithmic complexity theory addresses issues of contemporary concern including cryptography and data security, parallel computing, quantum computing, biological computing, circuit design, and the development of efficient algorithms (Homer and Selman, 2011). Lovász (1996) writes "complexity, I believe, should play a central role in the study of a large variety of phenomena, from computers to genetics to brain research to statistical mechanics. In fact, these mathematical ideas and tools may prove as important in the life sciences as the tools of classical mathematics (calculus and algebra) have proved in physics and chemistry" (pg. 1).

High School mathematics courses tend to be focused on preparing students for the study of calculus. Sylvia da Rosa (2004) claims that this has led students to the common misconception that mathematics is always continuous. In addition, few students are excited by pure mathematics. For most students to begin to appreciate mathematics, they have to see that it is useful. Incorporating the topic of algorithms in high school would afford teachers and students the opportunity to "apply the mathematics they know to solve problems arising in everyday life, society and the workplace" (CCSSI, 2010). Bernard Chazelle, a professor of computer science at Princeton University, in an interview in 2006 said on the future of computing: "The quantitative sciences of the 21st century such as proteomics and neurobiology, I predict, will place algorithms rather than formulas at their core. In a few decades we will have algorithms that will be considered as fundamental as, say, calculus is today."

Although there have been papers and books written on the history of algorithms, the history of algorithmic complexity has not been given nearly as much attention (Chabert, 1999; Cormen, et al., 2001). This paper will address this need by providing a history of algorithmic complexity, beginning with the text of Ibn al-majdi, a fourteenth century Egyptian astronomer, through the 21st century. In addition, this paper highlights the confusion surrounding big-O notation as well as the contributions of a group of mathematicians whose work in computability theory and complexity measures was critical to the development of the field of algorithmic complexity and the development of the theory of NP- Completeness (Knuth, 1976; Garey and Johnson, 1979). In effort to provide educators with context for which these topics can be presented, the problem of finding the maximum and minimum element in a sequence is introduced, along with an analysis of the complexity of two student-generated algorithms.

II. History of Algorithmic Complexity

Historically, an interest in optimizing arithmetic algorithms can be traced back to the Middle Ages (Chabert, 1999). Methods for reducing the number of separate elementary steps needed for calculation are described in an Arabic text by Ibn al-Majdi, a fourteenth century Egyptian astronomer. He compared the method of translation, which was used to find the product of two numbers, as well as the method of semi-translation, which was used only for calculating the squares of numbers. Based on Ibn al-Majdi's writings, if one were to use the method of translation to find the square of 348, for example, it would require nine multiplications. However, if one were to use the method of semi-translation to calculate the same product, it would require only six multiplications. In general, when squaring a number of n digits, the translation method takes n^2 elementary multiplications while the semi-translation method takes $n(n-1)/2$ elementary multiplications. In considering the number of steps, it is important to note that Ibn al-Majdi counted neither the number of elementary additions nor the doublings (Chabert, 1999).

Attempts to analyze the Euclidean algorithm date back to the early nineteenth century. In his well-known 1844 paper, Gabriel Lamé proved that if $u > v > 0$, then the number of division steps performed by the Euclidean algorithm $E(u,v)$ is always less than five times the number of decimal digits in v (Shallit, 1994). Although Lamé is generally recognized as the first to analyze the Euclidean algorithm, several other mathematicians had previously studied it. Shallit (1994) notes that sometime in between 1804 and 1811, Antoine-Andre-Louis Reynaud proved that the number of division steps performed by the

Euclidean algorithm, E(u,v), is less than or equal to v. Several years later he refined his upper limit to $v/2$ (which Lamé proved to be false).

Schreiber (1994) notes that algorithmic thinking has been a part of the study of geometric constructions since Antiquity. He explains that "the first studies on the unsolvability of problems by special instruments (i.e. by special classes of algorithms) and the first attempts to measure, compare, and optimize the complexity of different algorithms for solving the same problem" were in the field of geometric constructions (Schreiber, 1994, p. 691). These attempts can be seen in Émil Lemoine's 1902 'Géométrographie' (Schreiber, 1994).

In 1864, Charles Babbage predicted the significance of the study of algorithms. He wrote, "As soon as an Analytical Engine [i.e. general purpose computer] exists, it will necessarily guide the future course of the science. Whenever any result is sought by its aid, the question will then arise - By what course of calculation can these results be arrived at by the machine in the shortest time (Knuth, 1974, p. 329)?" The time taken to execute an algorithm, as Babbage predicted, is an important characteristic in quantifying an algorithm's efficiency.

In 1937, Arnold Scholtz studied the problem of optimizing the number of operations required to compute x^n (Kronsjö, 1987). In order to compute x^{31}, for example, it can be done in seven multiplications: $x^2, x^3, x^5, x^{10}, x^{20}, x^{30}, x^{31}$. However, if division is allowed, it can be done in six arithmetic operations.. Kronsjö (1987) notes that the problem of computing x^n with the fewest number of multiplications is far from being solved.

In the 1950's and 1960's, several mathematicians worked on optimization problems similar to that of Scholtz and Ibn al-Majdi. In order to evaluate a polynomial function $f(n) = a_n x^n + a_{n-1} x^{n-1} + ... + a_1 x + a_0$ at any point, it requires at most n additions and $2n$-1 multiplications. Horner's method, which involves rewriting the polynomial in a different form, only requires n multiplications and n additions, which is optimal in the number of multiplications under the assumption that no preprocessing of the coefficients is made (Kronsjö, 1987). In order to multiply two 2x2 matrices using the standard method, it requires eight multiplications and four additions[2]. In 1968, V. Strassen introduced a 'divide-and-conquer' method that reduced the number of multiplications to seven at the cost of fourteen more additions (Wilf, 2002). Generalizing his method to the multiplication of two nxn matrices (where n is an even number) reduces the number of multiplications from n^3 multiplications to $7n^3/8$ multiplications (Kronsjö, 1987). It is important to note that if additions are taken into consideration, Strassen's method is less efficient than the standard method for small inputs (Sedgewick, 1983). In order to compare different methods for solving a given problem, one needs to decide which operations should be counted and how they should be weighted. This decision, which is highly contingent on the method of implementation, is extremely important when comparing the efficiency of different algorithms. For example, computers typically consider multiplication to be more complex than addition. As a result, reducing the number of multiplications at the expense of some extra additions was preferable.

The introduction of the Turing machine in 1937 led to the development of the theory explaining which problems can and cannot be solved by a computer. This brought

---

[2] See Appendix for a more detailed comparison of the standard algorithm and Strassen's algorithm for matrix multiplication.

about questions regarding the relative computational difficulty of computable functions, which is the subject matter of computational complexity (Cook, 1987). Michael Rabin was one of the first to address what it means to say that a function f is more difficult to compute than a function g in his 1960 paper *Degree of Difficulty of Computing a Function and Hierarchy of Recursive Sets*. Juris Hartmanis and Richard Stearns, in their 1965 paper *On the Computational Complexity of Algorithms*, introduced the notion of complexity measure defined by the computation time on multitape Turing machines. Around the same time, Alan Cobham published *The Intrinsic Computational Difficulty of Functions*, which discussed machine-independent theory for measuring computational difficulty of algorithms. He considered questions such as whether multiplication is harder than addition and what constitutes a "step" in computation (Cook, 1987). These fundamental questions helped shape the concept of computational complexity. The field of computational complexity, which may be credited to the pioneering work of Stephen Cook, Richard Karp, Donald Knuth, and Michael Rabin, categorizes problems into classes based on the type of mathematical function that describes the best algorithm for each problem.

Several of the early authors on computational complexity struggled with the question of finding the most appropriate method to measure complexity. Although most agreed on computational time and space, the methods posed for these measurements varied. Kronsjö (1987) and Rosen (1999) define the time complexity of an algorithm as the amount of time used by a computer to solve a problem of a particular size using the algorithm. They define the space complexity of an algorithm as the amount of computer memory required to implement the algorithm. Thus, space and time complexity are tied to the particular data structures used to implement the algorithm. Although the concept of complexity is often addressed within the context of computer programs, given the variability in the space and speed of computers, a measurement that is independent of the method of implementation is often preferred.

A useful alternative, Kronsjö (1987) and Rosen (1999) note, is a mathematical analysis of the intrinsic difficulty of solving a problem computationally. They describe the computational complexity of an algorithm as the computational power required to solve the problem. This is measured by counting the number of elementary operations performed by the algorithm. The choice of elementary operation or operations will vary depending on the nature of the problem that the algorithm is designed to solve. It should however, be fundamental to the algorithm; for example, the number of real number multiplications and/or additions needed to evaluate a polynomial, the number of comparisons needed to sort a sequence, the number of multiplications needed to multiply two matrices. As long as the elementary operations are chosen well and are proportional to the total number of operations performed by the algorithm, this method will yield a consistent measurement of the computational difficulty of an algorithm which can be used to compare several algorithms for the same problem. Generally, by analyzing the complexity of several candidate algorithms for a problem, the most efficient one can be identified.

The number of elementary operations performed by an algorithm typically grows with the size of the input. Therefore, it is customary to describe the computational complexity or simply, complexity of an algorithm as a function of n, the size of its input (Cormen, et al., 2001). Additionally, the choice of n depends on the context of the problem for which the algorithm is being used. For sorting and searching problems, the most natural measure of n is the number of items in the input sequence. If the problem were to evaluate a polynomial, n would be better suited as the degree of the polynomial. If it were to multiply square matrices, on the other hand, n would represent the degree of the matrices. When n is sufficiently small, two algorithms may solve a problem using the same number of elementary operations. As n increases, one of the algorithms may perform significantly fewer elementary operations than the other, which would identify it as being more efficient. As such, it is important to consider the behavior of algorithms for large values of n. This can be done by comparing the growth rates of the complexity functions for each algorithm (to be discussed shortly).

For certain algorithms, even for inputs of the same size, the number of elementary operations performed by the algorithm can vary depending on the structure of the input. For example, an algorithm for alphabetizing a list of names may require a small number of comparisons if only a few of the names are out of order, and more comparisons if many of the names are out of order. We can classify algorithms according to whether or not their complexity depends on the structure of the input. An algorithm is said to be 'oblivious' if its complexity is independent of the structure of the input (Libeskind-Hadas, 1998). For 'non-oblivious' algorithms (whose complexity depends on the structure of the input) one must differentiate between the worst-case, average-case, and best-case scenarios by defining a separate complexity function for each case. For 'oblivious' algorithms it suffices to describe their complexity by a single function (as their worst-case, average-case, and best-case scenarios are all the same).

The worst-case complexity of an algorithm is the greatest number of operations needed to solve the problem over all inputs of size n. The best-case complexity of an algorithm is the least number of operations needed to solve the problem over all inputs of size n. The average-case complexity of an algorithm, is the average number of operations needed to solve the problem over all possible inputs of size n assuming all inputs are equally likely (Maurer and Ralston, 2004).

III. Big-O Notation

In order to compare the efficiencies of competing algorithms for a given problem, it is necessary to consider the number of operations performed by each algorithm for large inputs. This is done by classifying and comparing the growth rates of each algorithm's complexity function. Big-O notation, which was introduced by the German mathematician Paul Bachmann in 1894, is used extensively in the analysis of algorithms to describe the order of growth of a complexity function (Rosen, 1999). In particular, big-O gives an upper bound on the order of growth of a function.

Definition 1: "big-O": Let $f(n)$ and $g(n)$ be two positive valued functions, we say that $f(n) = O(g(n))$, if there is a constant c such that $f(n) \leq cg(n))$ for all but finitely many n.

If $f(n) =O(g(n))$ we say "$f(n)$ is $O(g(n))$." A geometric representation of $f(n) =O(g(n))$ can be seen in the figure below.

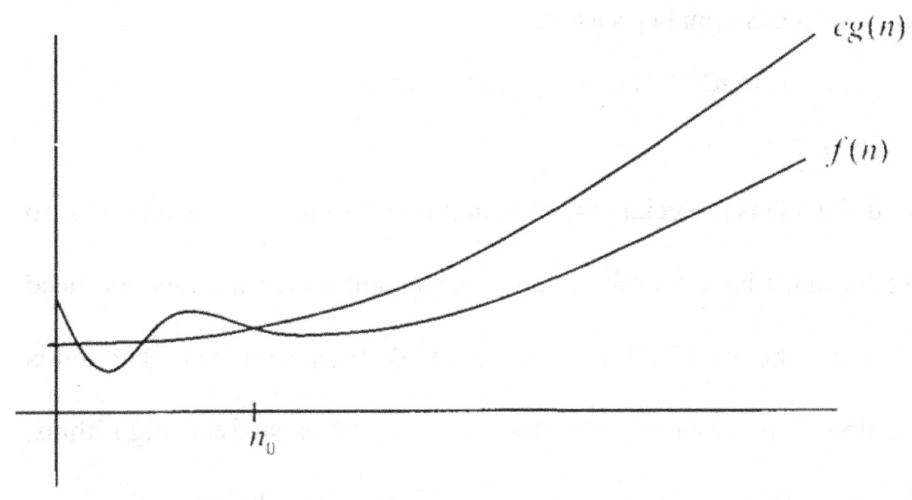

Figure 1

Knuth (1976) mentions that people often misuse big-O notation by assuming that it gives an exact order of growth; using it as if it specifies both a lower bound as well as an upper bound. Motivated by the misuse of big-O notation, Knuth introduced the big-Omega and big-Theta notations in the 1970s (Knuth, 1976). Big-Omega (denoted by the symbol Ω) provides a lower bound on the order of growth of a function, while big-Theta (denoted by the symbol Θ) provides both an upper and lower bound on the order of growth of a function. One algorithm is considered to be more efficient than another algorithm if its worst-case-complexity function has a lower order of growth. Unlike big-O and big-Omega, big-Theta gives an exact order without being precise about constant factors and is often the most appropriate of the three for comparing the efficiencies of different algorithms.

Definition 2 "big-Theta": Let $f(n)$ and $g(n)$ be two positive valued functions. We say that $f(n)=\Theta(g(n))$ if there is a constant $c\neq 0$ such that

$$\lim_{n\to\infty}\frac{f(n)}{g(n)}=c \qquad (1)$$

or if there are positive constants c_1 and c_2 such that

$$c_1 g(n) \leq f(n) \leq c_2 g(n) \qquad (2)$$

for all but finitely many n.

Note that condition (1) is a special case of condition (2) in that for condition (1) to be satisfied the function must have a limit, whereas for condition (2) the function need only be bounded above and below. Condition (1) is included because it is simpler and is sufficient to analyze almost all complexity functions that arise when studying algorithms. A geometric representation of $f(n) = \Theta(g(n))$ can be seen in the figure below.

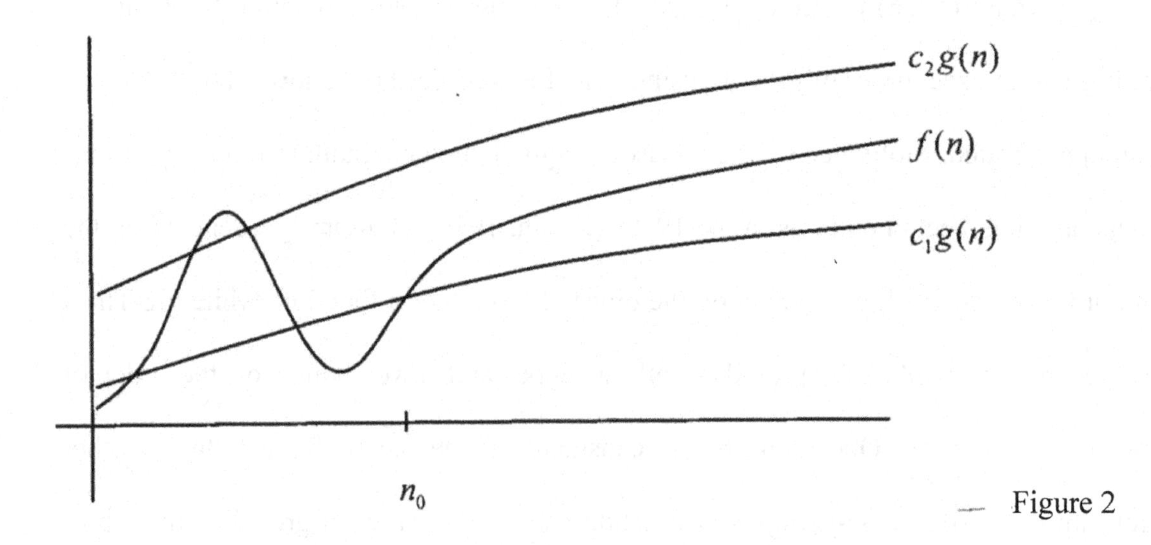

Figure 2

Definition 3 "big-Omega": Let $f(n)$ and $g(n)$ be two positive valued functions, we say that $f(n) = \Omega(g(n))$, if there is a constant c such that

$$f(n) \geq cg(n))$$

for all but finitely many n.

If $f(n) = \Omega(g(n))$ we say "$f(n)$ is $\Omega(g(n))$." A geometric representation of $f(n) = \Omega(g(n))$ can be seen in the figure below.

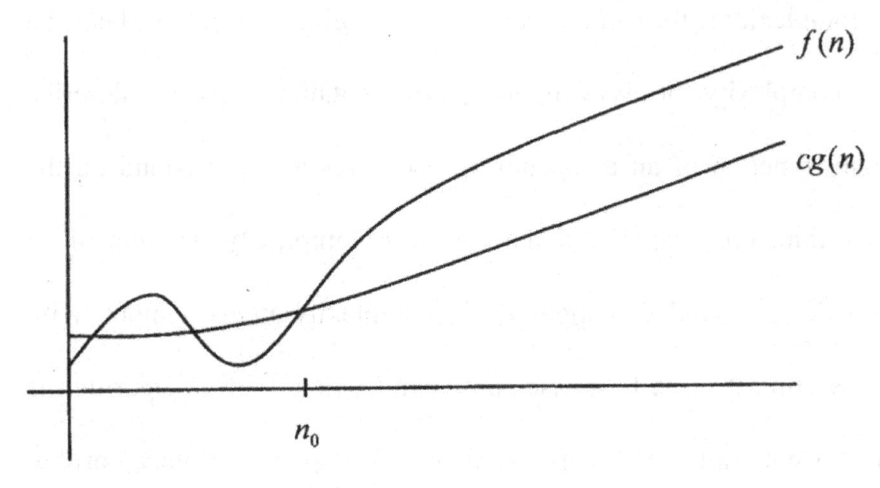

Figure 3

Of the three notations, big-O is the most widely used to classify the complexity of algorithms. In particular, an algorithm with an input size of n is said to have constant time complexity if its complexity is O(1), in other words, if its complexity is bounded above by a constant. Furthermore, an algorithm with an input size of n is said to have logarithmic time complexity if its complexity is O(log$n$), linear time complexity if its complexity is O(n), $n$log$n$ time complexity if its complexity is O($n$log$n$), polynomial time complexity if its complexity is O(n^k), exponential time complexity if its complexity is O(b^n) where $b>1$ and factorial time complexity if its complexity is O(n!).

An advantage of using big-O instead of big-Theta or big-Omega, is that when big-O is used to describe the worst-case complexity function of an algorithm, it also gives an upper bound on the complexity of the algorithm for *every* input. For example, if the worst-case complexity function of an algorithm is O($g(n)$), we can conclude that the complexity of the algorithm is O($g(n)$). On the other hand, if the worst-case complexity function of an algorithm is $\Theta(g(n))$, this does not does not imply an $\Theta(g(n))$ bound on the complexity of the algorithm for *every* input of size n. If however, the algorithm is 'oblivious' (in that its complexity is the same for all inputs of a given size), big-Theta can be used to describe the complexity for *every* input. When Ω-notation is used to describe the best-case complexity function of an algorithm, it also gives a lower bound on the complexity of the algorithm. For example, if the best-case complexity function of an algorithm is $\Omega(g(n))$ this lower bound also applies to its complexity on *every* input. With big-Theta, however, if we find that the best-case complexity function of an algorithm is $\Theta(g(n))$, this does not does not imply a Θ $(g(n))$ bound on the complexity of the algorithm for *every* input.

Maurer and Ralston (2004) and Knuth (1976) note that some textbooks mistakenly treat big-O and big-Theta as though they are the same. This has led to the misconception that algorithms can be compared by looking at what they are big-O of, which is not always the case. Suppose that one algorithm for a given problem has worst-case complexity function $f(n)=n$ and another algorithm for the same problem has worst-case complexity function $g(n)=n^2$. We can say that both $f(n)$ and $g(n)$ are $O(n^2)$. If we compared what the functions are big-O of, there is no indication as to which algorithm is more efficient. On the other hand if we considered using big-Theta to describe the growth of the complexity function we would find that $f(n)= \Theta(n)$ and $g(n)= \Theta(n^2)$. By using big-Theta we can observe that one algorithm (namely the one with worst-case complexity function $f(n)$) is more efficient than the other. As this example illustrates, the issue that arises in using big-O for comparing worst-case complexity functions lies in the fact that big-O notation only provides an upper bound which is not uniquely defined. Big-Theta, on the other hand, provides both an upper and lower bound. When we show that $f(n)= \Theta(g(n))$ it follows that $f(n)=O(g(n))$ but the converse may be false. In other words, if

$f(n)=O(g(n))$ it does not necessarily follow that $f(n)=\Theta(g(n))$. Hence if one wants to compare the worst-case complexity function of different algorithms to see which is more efficient, big-Theta would be the most appropriate. If, however, there is only an interest in providing an upper bound on the complexity of an algorithm, it suffices to describe its worst-case complexity function using big-O notation. Note that in the case of 'oblivious' algorithms where the complexity is the same for all inputs of size n, worst-case does not need to be specified.

In order to describe the complexity of an algorithm one can show that the worst-case complexity function of the algorithm is $O(g(n))$ or that worst-case complexity function of the algorithm is $\Theta(g(n))$. In either case it can be said that the complexity of the algorithm is $O(g(n))$). Note that the second case follows from the fact that if a function $f(n)$ is $\Theta(g(n))$ then $f(n)$ is $O(g(n))$.

IV. NP-Completeness

An important concept that was developed by 1965 was the identification of the class of problems solvable by algorithms with polynomial time complexity. The distinction between algorithms with polynomial and exponential time complexities was made as early as 1953 by Von Neumann, but the class of problems was not defined formally until Cobham in 1964 (Cook, 1987). Exponential time algorithms often perform exhaustive searches, whereas polynomial time algorithms rely on deeper insight into the structure of a problem. There is wide agreement that a problem has not been "well-solved" until a polynomial time algorithm is known for it (Garey and Johnson, 1979). A problem that is solvable using an algorithm with polynomial time complexity is called 'tractable' or 'easy', whereas problems that cannot be solved using an algorithm with polynomial time complexity are called ‘intractable’ or ‘hard’ (Rosen, 1999). The first examples of ‘intractable’ problems were obtained in the early 1960’s, as part of work on complexity “hierarchies” by Hartmanis and Stearns (Garey and Johnson, 1979). By the late 1960’s, a sizable class of practical problems that had not so far been solved with polynomial time algorithms was developed. These problems, which were known as NP-Complete, are believed to have the property that no algorithm with polynomial time complexity solves them, but that once a solution is known, it can be checked in polynomial time complexity. Additionally, these problems share the property that if any of them can be solved by an algorithm with polynomial time complexity, then they can all be solved by algorithms with polynomial time complexity (Rosen, 1999). To understand this difference, consider the problem of finding a solution to a Diophantine equation. There is no general method for finding a solution, however, it is relatively easy to check a proposed solution (Karp, 1987).

The foundations for the theory of NP-Completeness were laid in Cook’s 1971 paper, entitled *The Complexity Theorem Proving Procedures.* Subsequently, Karp presented a collection of results in his influential 1972 paper, which showed that twenty one important problems are NP-Complete. This generated tremendous interest in the notion of NP-Completeness. The question of whether or not the NP-Complete problems are ‘intractable’ is considered to be one of the foremost open questions of contemporary mathematics and computer science (Garey and Johnson, 1979). While no efficient algorithm for an NP-Complete problem has been found, it has yet to be proven that an efficient algorithm for one cannot exist.

Two well-known NP-Complete problems are the ‘Number Partitioning Problem’ and 'The Traveling Salesman Problem.' Suppose a group of athletes want to split up into two teams that are evenly matched. If the skill of each player is measured by an integer, can the athletes be split into two groups such that the sum of the skills in each group is the same? This is an example of the 'Number Partitioning Problem', a classic and surprisingly difficult problem in computer science, often called the 'easiest hard problem' (Hayes, 2002). The problem can be described as follows: given a set of n positive integers, separate them into two subsets such that the sum of the subsets is as close as possible to each other. Ideally, the two sums would be equal, but this is possible only if the sum of the entire set is even; in the event of an odd total, the best you can possibly do is to choose two subsets that differ by one. Try the problem on an arbitrary set of numbers such as $\{62, 24, 59, 71, 32, 25, 21, 39, 36, 63\}$. How many different possibilities are there? As you see, for large values of n this can become very time consuming! In order to find the subset pair whose sum is the closest, consider all possible subset pairs, calculate their sums, and return the pair whose sum is the closest. Since the number of subsets for an n element set is given by 2^n, as n increases the number of possibilities grows exponentially.

The 'Traveling Salesman Problem' can be described as follows: given n cities where n is a positive integer, and the distances between every pair of n cities, find a tour of minimal length, where a tour is a closed path that visits every city exactly once. Consider for example, the problem of finding the best tour of the state capitals of the United States. Provided the cost of traveling between cities is symmetric, the number of tours of n cities is $(n\text{-}1)!/2$. In order to find the best tour of the state capitals, this would require calculating $49!/2$ distances and then finding the shortest one. This would take even the fastest computers billions of years to solve (Papadimitriou and Steiglitz, 1982).

Although there are no efficient algorithms for finding the best possible solution to NP-Complete problems like the 'Number Partitioning Problem' and the 'Traveling Salesman Problem,' many approximation algorithms which provide good but not necessarily the best solution have been developed. Knowledge of NP-Complete problems is important because they arise surprisingly often in real-life applications and once a problem has been identified as NP-Complete, a lot of time can be saved by finding a good solution instead of trying to find the best possible solution (Cormen, et al., 2001).

V. A Teaching Example

This section will provide some insight on how the concept of algorithmic complexity can be incorporated into the existing high school or undergraduate mathematics curriculum. Teachers can start by presenting students with novel problems and encouraging them to develop their own algorithmic solutions. This can be more empowering to students than simply giving them an algorithm and asking them to analyze it. In particular, I have found students to be successful in developing their own algorithms for the 'Minimum/Maximum Problem.' In the pages that follow I will introduce two student-generated algorithms for this problem as well as an analysis of their respective complexities.

The 'Minimum/Maximum Problem' can be presented as follows:

Given a sequence of n distinct elements where each pair of elements can be ordered, find the minimum and maximum elements in the sequence. Students generally prefer using numbers to represent the elements in the sequence, though letters work just as well. In order to solve the problem of finding the minimum and maximum it may be helpful to work through the simpler problem of just finding the minimum. The only operation that can be used to gain information about the sequence is the comparison of two elements. Hence, we will consider the comparison of two elements to be an elementary operation. Let $f(n)$ represent the number of comparisons necessary to find the minimum element of a sequence of n elements (also known as a sequence of length n). For $n=2$, to find the minimum, we compare the two elements in the sequence, hence, $f(2)=1$. If we were to add an element to this sequence, we would compare the additional element to the minimum of the existing two elements; hence, for $n=3$, $f(3)=2$. In general, if we know the minimum element of a sequence of length k-1, then we could compare this minimum to the kth element to find the minimum of a sequence of length k. So for a sequence of length n, the number of comparisons is given by $f(n)=n-1$.

Next, let's consider the 'Minimum/Maximum Problem.' Students can be encouraged to pick a sequence of numbers and then consider the steps they would go through in order to find the minimum and maximum values. Students can formulate their algorithm by detailing the steps for a general sequence. They can then use algebra to analyze and describe the complexity of their algorithm. Some questions to pose to students to aid in their analysis are as follows:

(1) What "basic operation" is used to solve the problem?

(2) What properties of the input does the number of operations performed by the algorithm depend upon?

(3) Can you determine if your algorithm is 'oblivious' or 'non-oblivious'?

(4) Can you construct a complexity function to describe the number of operations performed by the algorithm for an input of a given size?

(5) If your algorithm is 'non-oblivious' can you construct a complexity function for the best and worst case scenarios?

In my experience with the 'Minimum/Maximum Problem,' high school mathematics students were able to come up with the following two algorithms rather quickly. The first algorithm uses the same approach as the method described earlier to find the minimum. Start by comparing the first two elements in the sequence to determine their minimum. Then compare the minimum to the third element to determine the minimum for the first three elements in the sequence. Continue this process until you've compared the last element in the sequence to the minimum of the others, for a total of n-1 comparisons. Once the minimum element of the entire sequence has been found, apply the same method to find the maximum, comparing elements to see which is greater. This would use an additional n-2 comparisons (as the minimum element need not be compared). Notice that this algorithm is 'oblivious,' in that the number of comparisons is the same for all inputs of size n. Let $g(n)$ represent the complexity function for this algorithm. Thus $g(n)=(n-1)+(n-2)=2n-3$ gives the total number of comparisons necessary to find the minimum and maximum of a sequence of length n. Using big-O notation we could classify this algorithm as having linear time complexity, or $O(n)$. However, since the algorithm is 'oblivious,' it can also be described as having complexity $\Theta(n)$.

Now let's consider a second, more complicated, student-generated algorithm. Start by comparing the first two elements in the sequence. Create an ordered pair with the two elements so that the smaller is in the first position and the larger is in the second position. Then compare the third element to the first term of the ordered pair. If it is smaller, let it take the place of the first term. Otherwise, compare it to the second term. If it is smaller than the second term, move on to the next element in the sequence. If it is larger than the second term, let it take the place of second term. Continue this process for the remaining elements in the sequence comparing them to the first and last term of the ordered pair until there are no more elements in the sequence to compare. The first term of the ordered pair will be the minimum of the sequence and the last term will be the maximum of the sequence.

As with the first algorithm, the only operation that can be used to gain information about the sequence is the comparison of two elements. Hence, we will consider the comparison of two elements to be an elementary operation. Can you see why this algorithm is more complicated than the other algorithm? This algorithm is 'non-oblivious.' The number of comparisons that it performs will depend on the size as well as the structure of the sequence. The structure is relevant to this algorithm because when each element is compared to the first term in the ordered pair, a decision is made whether or not to compare it to the second term of the ordered pair. As a result, the algorithm may perform one or two comparisons for each element beyond the first two elements. In the best-case scenario, the algorithm would use one comparison for the first two elements and then one comparison for each of the remaining n -2 elements. This gives a best-case complexity function of $b(n)= 1+ (n-2)= n - 1$. Can you think of what sequence structure would result in the best-case scenario? One example is when the sequence is in decreasing order. Now let's consider the worst-case scenario. In this case, the algorithm would use one comparison for the first two elements and then two comparisons for each of the remaining n-2 elements. This gives a worst-case complexity function of $w(n)= 1+ 2(n-2)=2n - 3$. Can you think of what sequence structure would result in the worst-case scenario? One example is when the sequence is in increasing order. Since the worst-case complexity functions is linear we can describe this algorithm as having linear time complexity, or $O(n)$. However since the best-case complexity function is linear as well, we could also describe this algorithm as being $\Omega(n)$. Furthermore, since the algorithm is both $O(n)$ and $\Omega(n)$, it follows that complexity is also $\Theta(n)$!

Although big-O, Ω, and Θ are widely used to compare complexity functions, for the two algorithms presented, the notation is not helpful in determining which is more efficient. This is due to the fact that they both have a complexity of $\Theta(n)$. We can however, perform a comparison of the actual complexity functions *g, b,* and *w.* Graph the three functions on the same axes. Notice that for a sequence of length $n=2$, both algorithms would perform only one comparison. As *n* increases, for some inputs the second algorithm will use fewer comparisons than the first algorithm. Therefore we can conclude that the second algorithm is more efficient. Now you might be wondering why we haven't looked at the average-case complexity of the second algorithm. Oftentimes, the average-case complexity function is much harder to describe. Students may assume that the average-case complexity function is $a(n)= 1.5n-2$ but further investigation will show that this is not the case.

VI. Conclusion

Algorithms are used to solve many of the problems that govern our everyday lives. In reviewing the history of algorithmic complexity we can see how early thinkers beyond developing the algorithms to solve problems, were interested in evaluating their techniques and working towards more efficient solutions. Knowledge of whether a problem can be solved efficiently is fundamental to solving the problem. Although computing power has increased in the last few decades, the amount of data that we are processing as a society has also grown tremendously. As a result, the ability to identify efficient algorithms for solving certain types of problems is gaining in importance (Homer and Selman, 2011). When reading texts on the analysis of algorithms it is helpful to understand the difference between the notations used to measure complexity and to be aware of the confusion that is prevalent in the literature. To better appreciate efficient algorithms we consider the alternative, algorithms whose complexity functions grow exponentially which the field of NP-Completeness encompasses. Furthermore, the 'Minimum/Maximum Problem' is included in order to provide a rich example of how algorithmic complexity can be analyzed at a level that is appropriate for high school or undergraduate mathematics students.

References

Capáy, M., and Magdin, M. (2013) Alternative Methods of Teaching Algorithms. *Social and Behavioral Sciences,* 83, 431 – 436.

Chabert, J. L. (Ed.) (1999). *A History of Algorithms: From the Pebble to the Microchip.* Berlin: Springer.

Common Core State Standards Initiative (CCSSI). (2010d) *Common Core State Standards for Mathematics*. Retrieved from http://www.corestandards.org/Math/.

Cook, S. A. (1987). An Overview of Computational Complexity. *ACM Turing Award Lectures: The First Twenty Years 1966-1985.* ACM Press Anthology Series.

Cormen, T., Leiserson, C., Rivest, R., and Stein, C. (2001). *Introduction to Algorithms* (2nd Edition). Massachusetts: The MIT Press.

da Rosa, S. (2004). Designing Algorithms in High School Mathematics. *Lecture Notes in Computer Science*, 3294, 17-31.

Garey, M. R. and Johnson, D. C. (1979). *Computers and Intractability: A Guide to the Theory of NP-Completeness.* San Francisco: W.H. Freeman and Company.

Hayes, B. (2002). The Easiest Hard Problem. *American Scientist*, 90(2), 113-117.

Homer, S. and Selman, A. (2011). *Computability and Complexity Theory* (2nd Edition). New York: Springer.

Karp, R. M. (1987). Combinatorics, Complexity, and Randomness, In *ACM Turing Award Lectures: The First Twenty Years 1966-1985* (pp. 433-453). New York: ACM Press Anthology Series.

Knuth, D. E. (1974). Computer Science and its Relation to Mathematics. *The American Mathematical Monthly*, 81, 323-343.

Knuth, D. E. (1976). Big Omicron and Big Omega and Big Theta. *ACM SIGACT News*, 8(2), 18-24.

Kronsjö, L. (1987). *Algorithms: Their Complexity and Efficiency* (2nd Edition). Great Britain: John Wiley and Sons.

Libeskin-Hadas, R. (1998) Sorting in Parallel. *The American Mathematical Monthly,* 105(3), 238-245.

Lovász, L. (1996). Information and Complexity (How To Measure Them?). In B. Pullman (Ed.), *The Emergence of Complexity in Mathematics, Physics, Chemistry and Biology, Pontifical Academy of Sciences* (pp. 65-80), Vatican City: Princeton University Press.

Maurer, S. B. and Ralston, A. (2004). *Discrete Algorithmic Mathematics* (3rd Edition). Massachusetts: A K Peters, Ltd.

Papadimitriou, C. H. and Steiglitz, K. (1982). *Combinatorial Optimization: Algorithms and Complexity*. New Jersey: Prentice-Hall, Inc.

Rosen, K. H. (1999). *Discrete Mathematics and its Applications*. United States: WCB McGraw-Hill.

Schreiber, P. (1994). *Algorithms and Algorithmic Thinking Through the Ages*. In I. Grattan-Guinness (Ed.), Companion Encyclopedia of the History and Philosophy of the Mathematical Sciences. New York: Routledge.

Sedgewick, R. (1983). *Algorithms*. Massachusetts: Addison-Wesley Publishing Company.

Shallit, J. (1994). Origins of the Analysis of the Euclidean Algorithm. *Historia Mathematica*, 21, 401-419.

Wilf, H. S. (2002). *Algorithms and Complexity*. Massachusetts: A K Peters, Ltd.

Appendix

Standard Algorithm:

Suppose you have two 2x2 matrices A=$\begin{bmatrix} a & b \\ c & d \end{bmatrix}$ and B=$\begin{bmatrix} e & f \\ g & h \end{bmatrix}$. Then the product matrix C=AB is given by C=$\begin{bmatrix} ae+bg & af+bh \\ ce+dg & cf+dh \end{bmatrix}$. An inspection of this example shows that at most 8 multiplications and 4 additions are necessary to multiply two 2x2 matrices. More generally to multiply two nxn matrices, computing each entry of the product matrix uses n multiplications and n-1 additions. Hence to compute the n^2 entries in the product matrix needs at most $n(n^2)=n^3$ multiplications and most $(n-1)(n^2)=n^3-n^2$ additions.

Strassen's Divide and Conquer Algorithm:

To multiply two 2x2 matrices A=$\begin{bmatrix} a & b \\ c & d \end{bmatrix}$ and B=$\begin{bmatrix} e & f \\ g & h \end{bmatrix}$, Strassen's algorithm first computes the following seven quantities, each of which requires exactly one multiplication:

$X_1=(a+b)(e+h)$

$X_2=(b-d)(g+h)$

$X_3=(a-c)(e+f)$

$X_4=(a+b)h$

$X_5=(c+d)e$

$X_6=a(f-h)$

$X_7=d(-e+g)$

Then the entries of the product matrix C=AB are computed as follows:

$ae + bg = X_1 + X_2 - X_4 + X_7$

$af + bh = X_4 + X_6$

$ce + dg = X_5 + X_7$

$cf + dh = X_1 - X_3 - X_5 + X_6$

To multiply two 2x2 matrices, Strassen's algorithm uses 7 multiplications and 18 additions, which reduces the number of multiplications used by the standard algorithm by 1 at the cost of 14 more additions.

Strassen's algorithm can be used to multiply larger matrices as well. Suppose we have two *n*x*n* matrices A and B where *n* is a power of 2. We partition A and B into four *n/*2x*n/*2 matrices and then multiply the parts recursively by computing the seven quantities defined above. Using Strassen's method uses $7n^3/8$ multiplications (Kronsjö, 1987). Manber (1988) notes that empirical studies indicate that *n* needs to be at least 100 to make Strassen's algorithm faster than the standard algorithm.

Integral of radical trigonometric functions revisited

Natanael Karjanto
Sungkyunkwan University, Republic of Korea

Binur Yermukanova
Nazarbayev University, Kazakhstan

ABSTRACT: This article revisits an integral of radical trigonometric functions. It presents several methods of integration where the integrand takes the form $\sqrt{1 \pm \sin x}$ or $\sqrt{1 \pm \cos x}$. The integral has applications in Calculus where it appears as the length of cardioid represented in polar coordinates.

Keywords: Techniques of integration, radical trigonometric functions, cardioid

The Mathematics Enthusiast, **ISSN 1551-3440, vol. 13, no. 3**, pp. 243–254

Introduction

This article revisits an integral where the integrand takes the form of radical trigonometric functions. A general form of radical trigonometric integrands in the context of this article refers to $\sqrt{a \pm b \sin x}$ or $\sqrt{a \pm b \cos x}$, for $a, b > 0$. The integral of these functions is expressed in terms of elliptic integral and are available in mathematical handbooks and tables of integrals. For example, the latter integral is given in Section 2.5 (see 2.576) of a famous mathematical handbook by Gradstyen and Ryzhik [GR]. For a particular case of $a = b$, after removing the constant factor, the integrand reduces to radical trigonometric functions $\sqrt{1 \pm \sin x}$ or $\sqrt{1 \pm \cos x}$. Interestingly, it seems that explicit expressions for the integral of these functions have not been specifically listed in any tables of integrals and handbooks, including, but not limited to, [AS, BSM, GR, Tab, Mat, PC, SLL]. The focus of this article is to consider the special case when $a = 1 = b$, where several techniques of integration are discussed in a more detail. To the best of our knowledge, this is the first time when such a compilation for particular integrands is presented.

The motivation of this article springs from an encounter from one of the coauthors' in teaching Calculus 2 course during the Spring 2014 semester in Nazarbayev University, Astana, Kazakhstan. Particularly, the content of this article is related to the topic on the integral calculus of polar curves, and one of the examples is calculating the length of a cardioid. We adopt the Calculus textbook written by Anton, Bivens and Davis [ABD] where the polar curves are discussed in Section 10.3. Another recommended textbook reading for this course is the one written by Stewart [Ste2]. An example [Example 4, Section 10.4, page 692] from the latter textbook mentions that finding the length of cardioid $r = 1 + \sin\theta$ can be evaluated by multiplying both the numerator and the denominator of the integrand by $\sqrt{2 - 2\sin\theta}$ or alternatively, using the Computer Algebra System (CAS). Yet, evaluating this integral by hand is apparently not so obvious to many students since they have to do further manipulation on the obtained expression. The screenshot of the example from Stewart's textbook has been excerpted and displayed in Figure 1.

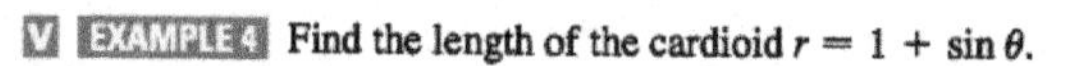
V EXAMPLE 4 Find the length of the cardioid $r = 1 + \sin\theta$.

SOLUTION The cardioid is shown in Figure 8. (We sketched it in Example 7 in Section 10.3.) Its full length is given by the parameter interval $0 \leqslant \theta \leqslant 2\pi$, so Formula 5 gives

$$L = \int_0^{2\pi} \sqrt{r^2 + \left(\frac{dr}{d\theta}\right)^2}\, d\theta = \int_0^{2\pi} \sqrt{(1 + \sin\theta)^2 + \cos^2\theta}\, d\theta$$

$$= \int_0^{2\pi} \sqrt{2 + 2\sin\theta}\, d\theta$$

We could evaluate this integral by multiplying and dividing the integrand by $\sqrt{2 - 2\sin\theta}$, or we could use a computer algebra system. In any event, we find that the length of the cardioid is $L = 8$.

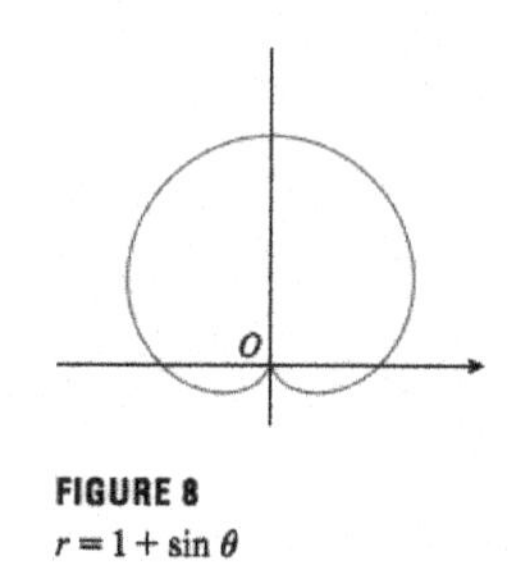

FIGURE 8
$r = 1 + \sin\theta$

Figure 1: An example from a textbook on calculating the length of cardioid $r = (1 + \sin\theta)$ where the calculation details are omitted.

After rationalizing the numerator and implementing the Pythagorean trigonometric identity, the numerator simplifies to $\sqrt{\cos^2\theta} = |\cos\theta|$, but it has to be in absolute value form, instead of simply $\cos\theta$. This is a common mistake found among students since they may forget or tend to ignore the absolute value sign. From an instructor's perspective, it is imperative to remind the students to be aware of this fact. Referring to Bloom's taxonomy of learning domains [Blo], the educational activity of this learning process is the *cognitive domain*. The process covers *knowledge*, *comprehension* and *application*. In this example, students possess the knowledge that any value of the square root must be non-negative and an absolute value of any quantity is always non-negative too. A comprehension of these facts is essential to conclude (application of knowledge) that the square root of a quantity squared is indeed equal to the absolute value of that quantity.

Referring to the revised Bloom taxonomy [AKB], a connection between learning activities and learning objectives can further be established. The knowledge dimension covers the *factual* and *conceptual* aspects. In this context, students must know the definition of an absolute value and be able to make an interrelationship between the property of a square root and the absolute value. The cognitive dimension includes *remember, understand, apply* aspects. Possessing the knowledge of absolute value, it is crucial to investigate whether the students can retrieve this knowledge from their memory, whether they understand why absolute value has to be non-negative and whether they are able to simplify and conclude that $\sqrt{\cos^2\theta} = |\cos\theta|$.

Another educational aspect of the integral involving radical trigonometric functions is related to the *synthesis* skill of *cognitive domain* in Bloom's taxonomy. In the revised Bloom's taxonomy, the educational content involves *factual* and *conceptual* aspects of the *knowledge* dimension, where students attempt to make interrelationships among the basic elements of trigonometric functions. The *cognitive process* dimension covers *remember, understand, apply* and *analyze* aspects. When this example is posed to the classroom for the students to work on, it turns out that some excellent students come up with different techniques by manipulating the integrand expression. This shows that different students approach the problem distinctly, they attempt to integrate with the method which is most convenient to them. For instance, for some students, the technique of trigonometric substitution is a more comfortable approach, others implement a variable shift method to solve the problem successfully. Thus, there are several ways by which students can approach the problem.

For many Calculus instructors, however, the interest in integration techniques has waned. With the introduction of CAS, many of them now give only cursory attention to such techniques. Nevertheless, the methods of integration covered in this article are still interesting from educational perspective. They provide a valuable pedagogical tool to assist and improve the students' learning skills, which are beneficial to both the instructors and the students themselves alike. In particular, by introducing several methods during class sessions, the techniques covered in this article become useful in the sense that it does not only expose the students to various techniques of integration but also makes them review and strengthen their knowledge of trigonometry and trigonometric functions. As can be observed later, this article recalls some important properties of trigonometric functions of sine, cosine and tangent as well as a significant application of trigonometric substitution in solving particular types of integration.

This article is organized as follows. The following section covers the integral of radical sine function $\sqrt{1 \pm \sin x}$. Section 2 briefly covers the integral of radical cosine function $\sqrt{1 \pm \cos x}$. Several techniques of integration are covered and more detailed derivations are discussed in Section 1, including rationalizing numerator, combining trigonometric identities, twice trigonometric substitutions and variable shift methods. All of these methods require some variations of integrating absolute value function, which will be presented accordingly in the corresponding subsections. Section 3 presents an application where the integrals of radical sine and cosine functions appear, particularly in calculating the length of a cardioid. The final section draws conclusions and provides remark to our discussion.

1 Integral of radical sine function

This section deals with the integral of a radical sine function where the integrand takes the form $\sqrt{1 \pm \sin x}$. There are a number of methods to obtain the result, and four techniques are covered in this section. The first method is by rationalizing the numerator. From here, one may depart either to use the Pythagorean identity or to employ a trigonometric substitution. The second method is by combining several trigonometric identities. We observe that double-angle formula and the identity relating $\sin x$ and $\tan x/2$ follow different paths of calculation and yet arrive at identical expression. The third technique is by implementing trigonometric substitutions two times, mainly using tangent function. Finally, the fourth technique is conducted by shifting the variable by $\pi/2$. Two options can be developed from this path, where both of them alter the integral from radical sine function into radical cosine function. The four methods covered in this section are summarized in the following tree diagram.

1.1 Rationalizing numerator

The following integral will be used in this subsection. Let f be a function which has at most one root on each interval on which it is defined, and F an antiderivative of f, i.e. $F'(x) = f(x)$, then

$$\int \frac{|f(x)|}{\sqrt{F(x)}}\,dx = -2\,\mathrm{sgn}[f(x)]\,\sqrt{F(x)} + C \tag{1.1}$$

where $\operatorname{sgn}(x)$ is the sign function, which takes the values $-1, 0$ or 1 when x is negative, zero or positive, respectively.

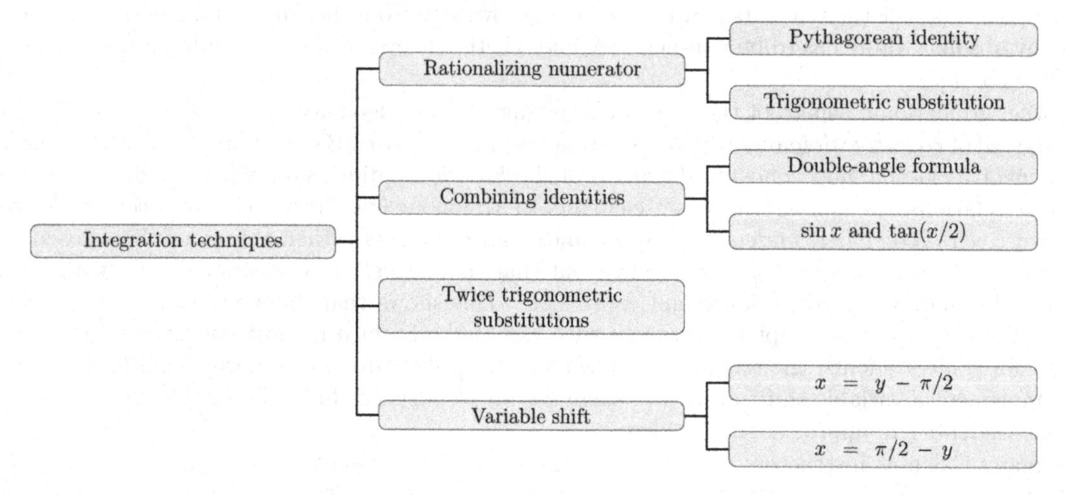

Pythagorean identity

Let I be an indefinite integral of the radical sine function $I = \displaystyle\int \sqrt{1 \pm \sin x}\, dx$, then rationalizing the numerator by multiplying both the numerator and the denominator with $\sqrt{1 \mp \sin x}$, applying the Pythagorean trigonometric identity and utilizing the definition of the absolute value, it yields:

$$\begin{aligned} I &= \int \sqrt{1 \pm \sin x} \cdot \frac{\sqrt{1 \mp \sin x}}{\sqrt{1 \mp \sin x}}\, dx = \int \frac{\sqrt{1 - \sin^2 x}}{\sqrt{1 \mp \sin x}}\, dx = \int \frac{\sqrt{\cos^2 x}}{\sqrt{1 \mp \sin x}}\, dx \\ &= \int \frac{|\cos x|}{\sqrt{1 \mp \sin x}}\, dx = -2\operatorname{sgn}(\cos x)\sqrt{1 \mp \sin x} + C \end{aligned}$$

where the last expression is readily obtained by implementing (1.1).

Trigonometric substitution

A similar solution can also be obtained using the trigonometric substitution of $u = \sin x$. Differentiating with respect to u, we get $dx = du/\cos x = du/(\pm\sqrt{1 - u^2})$, where the positive and negative signs are related to the sign of $\cos x$. Thus for $u = \sin x \neq \pm 1$

$$\begin{aligned} I &= \int \frac{\sqrt{1 \pm u}\, du}{\pm\sqrt{1 - u^2}} = \int \frac{du}{\pm\sqrt{1 \mp u}} = \mp 2\sqrt{1 \mp u} + C \\ &= \mp 2\sqrt{1 \mp \sin x} + C = -2\operatorname{sgn}(\cos x)\sqrt{1 \mp \sin x} + C. \end{aligned}$$

1.2 Combining identities

A general, explicit form of an integral involving an absolute value of a function will be used in this section. Let f be a function which has at most one root on each interval on which it is defined, and F an antiderivative of f that is zero at each root of f (such an antiderivative exists if and only if the condition on f is satisfied), then

$$\int |f(x)|\, dx = \operatorname{sgn}[f(x)]\, F(x) + C, \tag{1.2}$$

where $\operatorname{sgn}(x)$ is the sign function defined previously.

Double-angle formula

We manipulate the integrand by combining the Pythagorean trigonometric identity and the double-angle formula. Using the Pythagorean trigonometric identity, writing $1 = \cos^2(x/2) + \sin^2(x/2)$ and using the double-angle formula for $\sin x$: $\sin x = 2\sin(x/2)\cos(x/2)$, the integral of the radical sine becomes

$$\begin{aligned} I &= \int \sqrt{\cos^2\frac{x}{2} \pm 2\cos\frac{x}{2}\sin\frac{x}{2} + \sin^2\frac{x}{2}}\,dx = \int \sqrt{\left(\cos\frac{x}{2} \pm \sin\frac{x}{2}\right)^2}\,dx \\ &= \int \left|\cos\frac{x}{2} \pm \sin\frac{x}{2}\right|\,dx = 2\,\mathrm{sgn}\left(\cos\frac{x}{2} \pm \sin\frac{x}{2}\right)\left(\sin\frac{x}{2} \mp \cos\frac{x}{2}\right) + C \end{aligned}$$

where the last expression is quickly obtained after implementing (1.2).

Identity relating $\sin x$ and $\tan(x/2)$

A similar result will also be obtained if one employs another trigonometric identity that relates $\sin x$ and $\tan(x/2)$. Using the double-angle formula for $\sin x$ at the numerator and the Pythagorean trigonometric identity at the denominator, dividing both sides by $\cos^2(x/2)$, we obtain

$$\sin x = \frac{2\sin(x/2)\cos(x/2)}{\cos^2(x/2) + \sin^2(x/2)} = \frac{\frac{2\sin(x/2)\cos(x/2)}{\cos^2(x/2)}}{1 + \frac{\sin^2(x/2)}{\cos^2(x/2)}} = \frac{2\tan(x/2)}{1 + \tan^2(x/2)}.$$

Thus, the integral of the radical sine function I turns to

$$\begin{aligned} I &= \int \sqrt{1 \pm \frac{2\tan x/2}{1 + \tan^2 x/2}}\,dx = \int \sqrt{\frac{1 + \tan^2 x/2 \pm 2\tan x/2}{1 + \tan^2 x/2}}\,dx \\ &= \int \sqrt{\frac{(1 \pm \tan x/2)^2}{\sec^2 x/2}}\,dx = \int \left|\frac{1 \pm \tan x/2}{\sec x/2}\right|\,dx \\ &= \int \left|\cos\frac{x}{2} \pm \sin\frac{x}{2}\right|\,dx = 2\,\mathrm{sgn}\left(\cos\frac{x}{2} \pm \sin\frac{x}{2}\right)\left(\sin\frac{x}{2} \mp \cos\frac{x}{2}\right) + C. \end{aligned}$$

1.3 Twice trigonometric substitutions

A similar expression of the solution as that of the previous section can also be obtained by the trigonometric substitution $u = \tan x/2$. This implies $dx = 2\,du/(1+u^2)$ and writing $\sin x = 2\sin(x/2)\cos(x/2)$ the integral of the radical sine function becomes

$$\begin{aligned} I &= \int \sqrt{1 \pm \frac{2u}{1+u^2}}\,\frac{2du}{1+u^2} = 2\int \frac{|1 \pm u|\,du}{(1+u^2)^{3/2}} \\ &= 2\,\mathrm{sgn}(1 \pm u)\left(\int \frac{du}{(1+u^2)^{3/2}} \pm \int \frac{u\,du}{(1+u^2)^{3/2}}\right). \end{aligned}$$

Employ another trigonometric substitution $u = \tan y$ and $v = 1 + u^2$ for the first and the second integrals, respectively. Thus,

$$\begin{aligned} I &= 2\,\mathrm{sgn}(1 \pm u)\left(\int \frac{\sec^2 y\,dy}{(1+\tan^2 y)^{3/2}} \pm \frac{1}{2}\int \frac{dv}{v^{3/2}}\right) = 2\,\mathrm{sgn}(1 \pm u)\left(\int \frac{\sec^2 y\,dy}{\sec^3 y} \mp v^{-1/2}\right) \\ &= 2\,\mathrm{sgn}(1 \pm u)\left(\int \frac{1}{\sec y}\,dy \mp \frac{1}{\sqrt{v}}\right) = 2\,\mathrm{sgn}(1 \pm u)\left(\int \cos y\,dy \mp \frac{1}{\sqrt{1+u^2}}\right) \\ &= 2\,\mathrm{sgn}(1 \pm u)\left(\sin y \mp \frac{1}{\sqrt{1+u^2}}\right) + C = 2\,\mathrm{sgn}(1 \pm u)\,\mathrm{sgn}\left(\frac{1}{\sqrt{1+u^2}}\right)\left(\frac{u \mp 1}{\sqrt{1+u^2}}\right) + C \\ &= 2\,\mathrm{sgn}\left(1 \pm \tan\frac{x}{2}\right)\mathrm{sgn}\left(\cos\frac{x}{2}\right)\cos\frac{x}{2}\left(\tan\frac{x}{2} \mp 1\right) + C \\ &= 2\,\mathrm{sgn}\left(\cos\frac{x}{2} \pm \sin\frac{x}{2}\right)\left(\sin\frac{x}{2} \mp \cos\frac{x}{2}\right) + C. \end{aligned}$$

1.4 Variable shift

The following integrals of the absolute value of trigonometric functions $\sin \alpha x$ and $\cos \alpha x$, $\alpha \neq 0$, will be used in this subsection, where $\lfloor x \rfloor$ denotes the floor function:

$$\int |\sin \alpha x| \, dx = \frac{2}{\alpha} \left\lfloor \frac{\alpha x}{\pi} \right\rfloor - \frac{1}{\alpha} \cos\left(\alpha x - \left\lfloor \frac{\alpha x}{\pi} \right\rfloor \pi\right) + C \qquad (1.3)$$

$$\int |\cos \alpha x| \, dx = \frac{2}{\alpha} \left\lfloor \frac{\alpha x}{\pi} + \frac{1}{2} \right\rfloor + \frac{1}{\alpha} \sin\left(\alpha x - \left\lfloor \frac{\alpha x}{\pi} + \frac{1}{2} \right\rfloor \pi\right) + C. \qquad (1.4)$$

Variable shift $x = y - \pi/2$

Applying this variable shift, the integral I becomes

$$\begin{aligned}
I &= \int \sqrt{1 \pm \sin(y - \pi/2)} \, dy = \int \sqrt{1 \mp \cos y} \, dy \\
&= \begin{cases} \displaystyle\int \sqrt{2 \sin^2 y/2} \, dy, & \text{for } - \text{ sign } (+ \text{ sign original } I) \\ \displaystyle\int \sqrt{2 \cos^2 y/2} \, dy, & \text{for } + \text{ sign } (- \text{ sign original } I) \end{cases} \\
&= \begin{cases} \sqrt{2} \displaystyle\int |\sin y/2| \, dy, & \text{for } - \text{ sign } (+ \text{ sign original } I) \\ \sqrt{2} \displaystyle\int |\cos y/2| \, dy, & \text{for } + \text{ sign } (- \text{ sign original } I) \end{cases} \\
&= \begin{cases} -2\sqrt{2} \, \text{sgn}(\sin y/2) \cos(y/2) + C, & \text{for } - \text{ sign } (+ \text{ sign original } I) \\ 2\sqrt{2} \, \text{sgn}(\cos y/2) \sin(y/2) + C, & \text{for } + \text{ sign } (- \text{ sign original } I) \end{cases} \\
&= \begin{cases} -2\sqrt{2} \, \text{sgn}\left[\sin\left(\frac{x}{2} + \frac{\pi}{4}\right)\right] \cos\left(\frac{x}{2} - \frac{\pi}{4}\right) + C, & \text{for } - \text{ sign } (+ \text{ sign original } I) \\ 2\sqrt{2} \, \text{sgn}\left[\cos\left(\frac{x}{2} + \frac{\pi}{4}\right)\right] \sin\left(\frac{x}{2} + \frac{\pi}{4}\right) + C, & \text{for } + \text{ sign } (- \text{ sign original } I) \end{cases} \\
&= -2\sqrt{2} \, \text{sgn}\left[\sin\left(\frac{x}{2} \pm \frac{\pi}{4}\right)\right] \cos\left(\frac{x}{2} \pm \frac{\pi}{4}\right) + C
\end{aligned}$$

where the last three expressions are readily obtained by implementing (1.2), returning back the original variable and combining results corresponding to the positive and negative signs into a single expression, respectively. Alternatively, implementing (1.3), we obtain the integral for $\sqrt{1 + \sin x}$:

$$\begin{aligned}
I_1 &= 4\sqrt{2} \left\lfloor \frac{y}{2\pi} \right\rfloor - 2\sqrt{2} \cos\left(\frac{y}{2} - \left\lfloor \frac{y}{2\pi} \right\rfloor \pi\right) + C \\
&= 4\sqrt{2} \left\lfloor \frac{x}{2\pi} + \frac{1}{4} \right\rfloor - 2\sqrt{2} \cos\left(\frac{x}{2} + \frac{\pi}{4} - \left\lfloor \frac{x}{2\pi} + \frac{1}{4} \right\rfloor \pi\right) + C.
\end{aligned}$$

Implementing (1.4), we obtain the integral for $\sqrt{1 - \sin x}$:

$$\begin{aligned}
I_2 &= 4\sqrt{2} \left\lfloor \frac{y}{2\pi} + \frac{1}{2} \right\rfloor + 2\sqrt{2} \sin\left(\frac{y}{2} - \left\lfloor \frac{y}{2\pi} + \frac{1}{2} \right\rfloor \pi\right) + C \\
&= 4\sqrt{2} \left\lfloor \frac{x}{2\pi} + \frac{3}{4} \right\rfloor + 2\sqrt{2} \sin\left(\frac{x}{2} + \frac{\pi}{4} - \left\lfloor \frac{x}{2\pi} + \frac{3}{4} \right\rfloor \pi\right) + C
\end{aligned}$$

where subscripts 1 and 2 correspond to the positive and negative signs in the original integral I, respectively.

Variable shift $x = \pi/2 - y$

Applying this variable shift, the integral I becomes

$$\begin{aligned}
I &= -\int \sqrt{1 \pm \sin(\pi/2 - y)}\, dy = -\int \sqrt{1 \pm \cos y}\, dy \\
&= \begin{cases} -\sqrt{2}\displaystyle\int |\cos y/2|\, dy, & \text{for } + \text{ sign} \\ -\sqrt{2}\displaystyle\int |\sin y/2|\, dy, & \text{for } - \text{ sign} \end{cases} \\
&= \begin{cases} -2\sqrt{2}\,\text{sgn}(\cos y/2)\sin(y/2) + C, & \text{for } + \text{ sign} \\ 2\sqrt{2}\,\text{sgn}(\sin y/2)\cos(y/2) + C, & \text{for } - \text{ sign} \end{cases} \\
&= \begin{cases} 2\sqrt{2}\,\text{sgn}\left[\cos\left(\frac{x}{2} - \frac{\pi}{4}\right)\right]\sin\left(\frac{x}{2} - \frac{\pi}{4}\right) + C, & \text{for } + \text{ sign} \\ -2\sqrt{2}\,\text{sgn}\left[\sin\left(\frac{x}{2} - \frac{\pi}{4}\right)\right]\cos\left(\frac{x}{2} - \frac{\pi}{4}\right) + C, & \text{for } - \text{ sign} \end{cases} \\
&= 2\sqrt{2}\,\text{sgn}\left[\cos\left(\frac{x}{2} \mp \frac{\pi}{4}\right)\right]\sin\left(\frac{x}{2} \mp \frac{\pi}{4}\right) + C
\end{aligned}$$

where the last three expressions are readily obtained by implementing (1.2), returning back the original variable and combining two results into a single expression, respectively. Alternatively, implementing (1.4), we obtain the integral for $\sqrt{1 + \sin x}$:

$$\begin{aligned}
I_1 &= -4\sqrt{2}\left\lfloor \frac{y}{2\pi} + \frac{1}{2} \right\rfloor - 2\sqrt{2}\sin\left(\frac{y}{2} - \left\lfloor \frac{y}{2\pi} + \frac{1}{2} \right\rfloor \pi\right) + C \\
&= -4\sqrt{2}\left\lfloor \frac{3}{4} - \frac{x}{2\pi} \right\rfloor + 2\sqrt{2}\sin\left(\frac{x}{2} - \frac{\pi}{4} + \left\lfloor \frac{3}{4} - \frac{x}{2\pi} \right\rfloor \pi\right) + C \\
&= 4\sqrt{2}\left\lceil \frac{x}{2\pi} - \frac{3}{4} \right\rceil + 2\sqrt{2}\sin\left(\frac{x}{2} - \frac{\pi}{4} - \left\lceil \frac{x}{2\pi} - \frac{3}{4} \right\rceil \pi\right) + C
\end{aligned}$$

where $\lceil x \rceil$ is the ceiling function and the relationship between the floor and the ceiling functions are utilized to obtain the last expression, i.e. $\lfloor x \rfloor + \lceil -x \rceil = 0$. Implementing (1.3), we obtain the integral for $\sqrt{1 - \sin x}$:

$$\begin{aligned}
I_2 &= -4\sqrt{2}\left\lfloor \frac{y}{2\pi} \right\rfloor + 2\sqrt{2}\cos\left(\frac{y}{2} - \left\lfloor \frac{y}{2\pi} \right\rfloor \pi\right) + C \\
&= -4\sqrt{2}\left\lfloor \frac{1}{4} - \frac{x}{2\pi} \right\rfloor + 2\sqrt{2}\cos\left(\frac{x}{2} - \frac{\pi}{4} + \left\lfloor \frac{1}{4} - \frac{x}{2\pi} \right\rfloor \pi\right) + C \\
&= 4\sqrt{2}\left\lceil \frac{x}{2\pi} - \frac{1}{4} \right\rceil + 2\sqrt{2}\cos\left(\frac{x}{2} - \frac{\pi}{4} - \left\lceil \frac{x}{2\pi} - \frac{1}{4} \right\rceil \pi\right) + C
\end{aligned}$$

where the subscripts 1 and 2 correspond to the positive and negative signs in the expressions of I, respectively.

2 Integral of radical cosine function

This section compiles a number of techniques to integrate the radical cosine function in the form $\sqrt{1 \pm \cos x}$. Let J be an indefinite integral of radical cosine function $J = \int \sqrt{1 \pm \cos x}\, dx$. Since the derivations are similar to the ones in Section 1, only the final results will be presented. Employing the variable shift method either by $x = \pi/2 - y$ or $x = y - \pi/2$ will alter the cosine function into the sine function and vice versa. Thanks to this redundancy, the coverage of this technique will be omitted in this section. The integration techniques presented in this section basically can also be summarized with a similar tree diagram presented in Section 1.

2.1 Rationalizing numerator

Implementing two techniques of rationalizing numerator and by trigonometric substitution $u = \cos x$, we obtain a similar result to the one in the previous section:

$$J = -2\,\text{sgn}(\sin x)\sqrt{1 \mp \cos x} + C.$$

2.2 Combining identities

This technique deals with combining the Pythagorean trigonometric identity with the double-angle formula and the identity of $\cos x$ and $\tan(x/2)$. The double-angle formula used here is $\cos x = \cos^2(x/2) - \sin^2(x/2)$. The identity of $\cos x$ in terms of $\tan(x/2)$ reads

$$\cos x = \frac{1 - \tan^2(x/2)}{1 + \tan^2(x/2)}.$$

Employing these identities the integral J now reads

$$J = \begin{cases} 2\sqrt{2}\,\mathrm{sgn}\,[\cos(x/2)]\,\sin(x/2) + C, & \text{for } + \text{ sign} \\ -2\sqrt{2}\,\mathrm{sgn}\,[\sin(x/2)]\,\cos(x/2) + C, & \text{for } - \text{ sign.} \end{cases} \tag{2.1}$$

2.3 Twice trigonometric substitutions

Employing the substitution $u = \tan(x/2)$, we have

$$J = \begin{cases} \displaystyle\int \frac{2\sqrt{2}\,du}{(1+u^2)^{3/2}} = \mathrm{sgn}\left(\frac{1}{\sqrt{1+u^2}}\right)\frac{2\sqrt{2}u}{\sqrt{1+u^2}} + C, & \text{for } + \text{ sign} \\ \displaystyle\int \frac{2\sqrt{2}|u|\,du}{(1+u^2)^{3/2}} = \mathrm{sgn}\left(\frac{u}{\sqrt{1+u^2}}\right)\frac{-2\sqrt{2}}{\sqrt{1+u^2}} + C, & \text{for } - \text{ sign.} \end{cases}$$

After returning to the initial variable x, identical expressions with the ones in (2.1) will be obtained.

3 Application: Cardioid

The integral discussed above appears as calculation of the arc length of a cardioid. The length of cardioids $r = a(1 \pm \sin\theta)$, $a > 0$ is given by

$$L = a\sqrt{2}\int_0^{2\pi} \sqrt{1 \pm \sin\theta}\,d\theta.$$

The sketches of the cardioids are presented in Figure 2. The properties of the curve have been investigated in a classical paper by Yates more than half a century ago [Yat2]. The author also compiled a handbook on many kinds of curves, including cardioid, and discussed their properties [Yat1]. Another approach of calculating an area of cardioid and other shapes of closed curves is presented using the surveyor's method [Bra]. A road-wheel relationship by rolling a cardioid wheel on an inverted cycloid is discussed in [HW].

Cardioid finds various applications in fractals, complex analysis, plant physiology and engineering. In fractals, it appears in Douady cauliflower, which is a decoration formed via numerous small cardioids of the Mandelbrot set [PRÁ, RPÁ]. In plant physiology, the seed shape of *Arabidopsis* (rock cress) can be modelled using cardioid [CMA]. The model based on the comparison of the outline of the seed's longitudinal section with a transformed cardioid, where the horizontal axis is scaled by a factor equal to the Golden Ratio. An envelope of rays either reflected or refracted from the surface, known as caustic, from a cup of coffee or milk exhibits the shape of a cardioid [Cau]. In the field of electronics and electrical engineering, a cardioid directional pattern in a microphone provides a relatively wide pick-up zone [MH].

It is stated but not shown in a Calculus textbook authored by Stewart [Ste1, Ste2] that one can calculate this integral using the techniques described in this article or by technology, amongst others, are *Integral Calculator* [Cal], *Sage* [Sag], *Symbolab* [Sym] and *Wolfram Alpha* [Wol]. The author uses the cardioid $r = 1 + \sin\theta$ as an example, as shown in Figure 1 mentioned earlier in the introduction of this article. In general, evaluating a definite integral involving an absolute value, one must find the zeros of the function in the absolute value and divide the range of integration into pieces by toggling the sign within each of the intervals.

3.1 Rationalizing numerator

Since $\cos\theta \geq 0$ for $0 \leq \theta \leq \pi/2$ and $3\pi/2 \leq \theta \leq 2\pi$ and $\cos\theta < 0$ for $\pi/2 < \theta < 3\pi/2$, we need to split the integral into three intervals. See the top panel of Figure 3. Thus, using the result from Subsection 1.1,

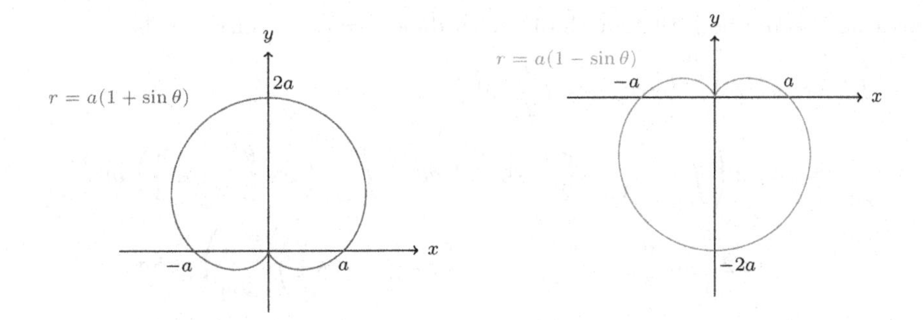

Figure 2: Sketches of cardioids $r = a(1 + \sin\theta)$ (left) and $r = a(1 - \sin\theta)$ (right), $a > 0$.

the length of the cardioids $r = a(1 \pm \sin\theta)$ is given by

$$\begin{aligned} L &= a\sqrt{2}\int_0^{2\pi} \frac{|\cos\theta|}{\sqrt{1 \mp \sin\theta}}\, d\theta \\ &= a\sqrt{2}\left(\int_0^{\pi/2} \frac{\cos\theta}{\sqrt{1 \mp \sin\theta}}\, d\theta - \int_{\pi/2}^{3\pi/2} \frac{\cos\theta}{\sqrt{1 \mp \sin\theta}}\, d\theta + \int_{3\pi/2}^{2\pi} \frac{\cos\theta}{\sqrt{1 \mp \sin\theta}}\, d\theta\right) \\ &= 2a\sqrt{2}\left(\mp\sqrt{1 \mp \sin\theta}\Big|_0^{\pi/2} \pm \sqrt{1 \mp \sin\theta}\Big|_{\pi/2}^{3\pi/2} \mp \sqrt{1 \mp \sin\theta}\Big|_{3\pi/2}^{2\pi} \right) = 8a. \end{aligned}$$

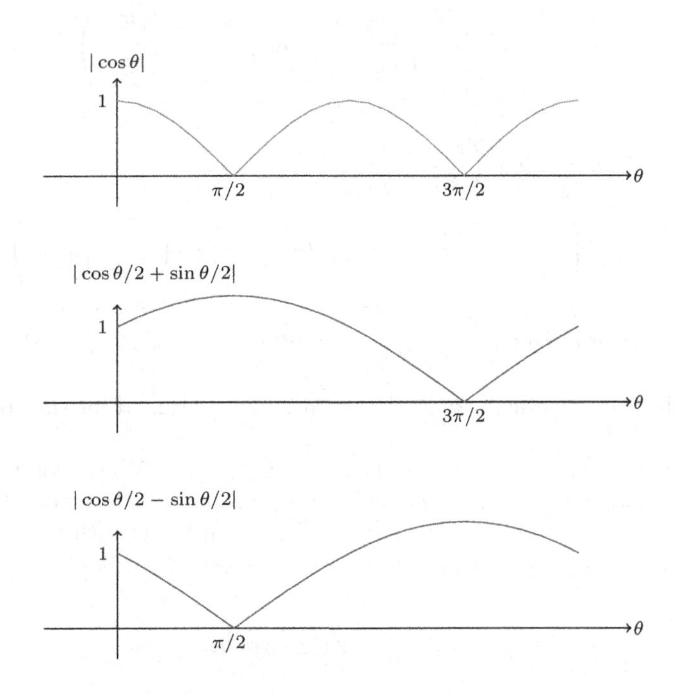

Figure 3: Plots of $|\cos\theta|$ (top panel), $|\cos(\theta/2) + \sin(\theta/2)|$ (middle panel) and $|\cos(\theta/2) - \sin(\theta/2)|$ (bottom panel) for $0 \le \theta \le 2\pi$ with the indicated zeros.

3.2 Twice trigonometric substitutions

We know that (see the middle panel of Figure 3)

$$\cos\frac{\theta}{2} + \sin\frac{\theta}{2} = \sqrt{2}\cos\left(\frac{\theta}{2} - \frac{\pi}{4}\right) \begin{cases} \ge 0, & \text{for} \quad 0 \le \theta \le 3\pi/2 \\ < 0, & \text{for} \quad 3\pi/2 < \theta \le 2\pi \end{cases}$$

Thus, implementing this method, the length of the cardioid $r = a(1 + \sin\theta)$ reads

$$\begin{aligned} L &= a\sqrt{2}\int_0^{2\pi}\left|\cos\frac{\theta}{2} + \sin\frac{\theta}{2}\right|\,d\theta \\ &= a\sqrt{2}\left(\int_0^{3\pi/2}\left(\cos\frac{\theta}{2} + \sin\frac{\theta}{2}\right)\,d\theta - \int_{3\pi/2}^{2\pi}\left(\cos\frac{\theta}{2} + \sin\frac{\theta}{2}\right)\,d\theta\right) \\ &= 2a\sqrt{2}\left(\sin\frac{\theta}{2} - \cos\frac{\theta}{2}\bigg|_0^{3\pi/2} - \left(\sin\frac{\theta}{2} - \cos\frac{\theta}{2}\right)\bigg|_{3\pi/2}^{2\pi}\right) = 8a. \end{aligned}$$

Similarly, splitting the integral at $\theta = \pi/2$, we also obtain the length $L = 8a$ corresponding to the cardioid $r = a(1 - \sin\theta)$. See the bottom panel of Figure 3 to observe that the zero of $\cos(\theta/2) - \sin(\theta/2)$ for $0 \le \theta \le 2\pi$ is at $\pi/2$.

3.3 Variable shift

These integrals involve the absolute value functions $|\cos(y/2)|$ and $|\sin(y/2)|$, for which in the original variable θ, both functions are non-negative for $0 \le \theta \le 3\pi/2$ and negative for $3\pi/2 < \theta < 2\pi$. Thus, the length of the cardioid $r = a(1 + \sin\theta)$ reads

$$\begin{aligned} L &= 2a\int_0^{2\pi}\left|\sin\left(\frac{\theta}{2} + \frac{\pi}{4}\right)\right|\,d\theta \\ &= 2a\left(\int_0^{3\pi/2}\sin\left(\frac{\theta}{2} + \frac{\pi}{4}\right)\,d\theta - \int_{3\pi/2}^{2\pi}\sin\left(\frac{\theta}{2} + \frac{\pi}{4}\right)\,d\theta\right) \\ &= 4a\left(-\cos\left(\frac{\theta}{2} + \frac{\pi}{4}\right)\bigg|_0^{3\pi/2} + \cos\left(\frac{\theta}{2} + \frac{\pi}{4}\right)\bigg|_{3\pi/2}^{2\pi}\right) = 8a \end{aligned}$$

or

$$\begin{aligned} L &= 2a\int_0^{2\pi}\left|\cos\left(\frac{\pi}{4} - \frac{\theta}{2}\right)\right|\,d\theta \\ &= 2a\left(\int_0^{3\pi/2}\cos\left(\frac{\pi}{4} - \frac{\theta}{2}\right)\,d\theta - \int_{3\pi/2}^{2\pi}\cos\left(\frac{\pi}{4} - \frac{\theta}{2}\right)\,d\theta\right) \\ &= 4a\left(-\sin\left(\frac{\pi}{4} - \frac{\theta}{2}\right)\bigg|_0^{3\pi/2} + \sin\left(\frac{\pi}{4} - \frac{\theta}{2}\right)\bigg|_{3\pi/2}^{2\pi}\right) = 8a. \end{aligned}$$

Employing a similar technique, identical result of $L = 8a$ is also obtained for the corresponding cardioid $r = a(1 - \sin\theta)$.

An expression $r = a(1 \pm \cos\theta)$, $a > 0$ produces cardioids too. When comparing this expression with the one with sine term, the effect is a 90-degree rotation, either clockwise (for the same sign) or counterclockwise (for the opposite sign), of the corresponding cardioids with the sine term. The sketch of the corresponding cardioid is presented in Figure 4. The length of cardioids $r = a(1 \pm \cos\theta)$, $a > 0$ is given by

$$L = a\sqrt{2}\int_0^{2\pi}\sqrt{1 \pm \cos\theta}\,d\theta.$$

Using similar techniques discussed in Section 2, one can find that the length of these cardioids is also $8a$.

A number of Calculus textbooks use this type of cardioid as an example for calculating its length. For instance, Anton et al. [ABD] uses the cardioid $r = 1 + \cos\theta$. After some manipulations, one needs to integrate $|\cos(\theta/2)|$ from $\theta = 0$ to $\theta = 2\pi$. Although general readers will attempt to split the boundary integrations at $\theta = \pi$, the authors explain that since the cardioid is symmetry about the polar axis, the integral from $\theta = \pi$ to $\theta = 2\pi$ is equal to the one from $\theta = 0$ to $\theta = \pi$. Thus, the integral can be calculated by twice integrating from $\theta = 0$ to $\theta = \pi$ of the positive integrand $\cos(\theta/2)$ (without the absolute value). Calculus' Thomas textbook [TWH] adopts the cardioid $r = 1 - \cos\theta$. The integrand reduces to $|\sin(\theta/2)|$. Fortunately, $\sin(\theta/2) \ge 0$ for $0 \le \theta \le 2\pi$ and thus by removing the absolute value and evaluating the integral, one can quickly obtain the length of the cardioid.

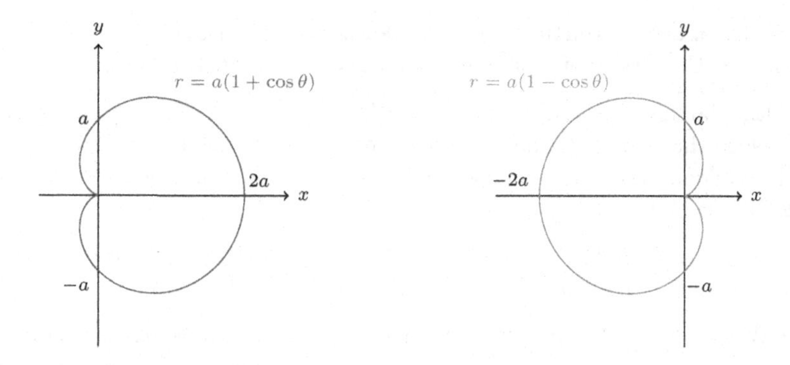

Figure 4: Sketches of cardioids $r = a(1 + \cos\theta)$ (left) and $r = a(1 - \cos\theta)$ (right), $a > 0$.

4 Conclusion and Remark

This article presents the integral with radical sine and cosine functions where its application appears in the length of a cardioid. It turns out that several techniques of integration exist to solve the problem, which is interesting from the perspective of teaching and learning mathematics. Despite the current trend of using CAS, the collection of integration techniques presented in this article is a valuable pedagogical tool. To the best of our knowledge, this is the first time such a compilation for this particular type of integrands is presented. We are convinced that this article contains useful educational contents that will be beneficial for both instructors and students alike. We also consider our contribution as a complement to existing Calculus textbooks which discuss a topic on calculating the length of a polar curve, particularly cardioid.

Acknowledgment. The authors wish to thank Dr. Ulrich Norbisrath (Faculty of Computer Science, Communication and Media, University of Applied Sciences Upper Austria), Dr. Richard J. Mathar (Max-Planck Institut für Astronomie, Heidelberg, Germany), Professor Victor Hugo Moll (Tulane University, New Orleans, Louisiana, USA), Professor Chris Sangwin (Mathematics Education Centre, Loughborough University and School of Mathematics, The University of Edinburgh, UK), the anonymous reviewers whose comments and remarks helped the improvement of this article, Murat Yessenov (Class 2017 of Physics major, SST, NU) and other students from SST and SHSS (School of Humanities and Social Sciences) who enrolled in Section 3 of MATH-162 Calculus 2 during Spring 2014 at NU, Astana, Kazakhstan.

References

[AS] Abramowitz, M., & Stegun, I. A. (1964). *Handbook of Mathematical Functions with Formulas, Graphs, and Mathematical Tables.* Eastford, Connecticut: Martino Fine Books.

[AKB] Anderson, L. W., Krathwohl, D. R., & Bloom, B. S. (2001). *A taxonomy for learning, teaching, and assessing: A revision of Bloom's taxonomy of educational objectives.* Boston, Massachusetts: Allyn & Bacon.

[ABD] Anton, H., Bivens, I. C., & Davis, S. (2012). *Calculus–Early Trancendentals*, 10th edition, (International Student version). Hoboken, New Jersey: John Wiley & Son's Inc.

[Blo] Bloom, B. S. (1956). *Taxonomy of educational objectives: The classification of educational goals. Handbook I: Cognitive domain.* New York: David McKay Company.

[Bra] Braden, B. (1986). The surveyor's area formula. *The College Mathematics Journal*, 17(4), 326–337.

[BCG] Briggs, W. L., Cochran, L., & Gillett, B. (2012). *Calculus for Scientists and Engineers.* Boston, Massachusetts: Pearson Education, Inc.

[BSM] Bronshtein, I. N., Semendyayev, K. A., Musiol, G., & Mühlig, H. (2007). *Handbook of Mathematics*, 5th edition. Berlin Heidelberg: Springer-Verlag.

[Cal] `http://www.integral-calculator.com/`. Developed by David Scherfgen from Hochschule Bonn-Rhein-Sieg University of Applied Sciences, Sankt Augustin, Germany.

[CMA] Cervantes, E., Martín, J. J., Ardanuy, R., de Diego, J. G., & Tocino, Á. (2010). Modeling the Arabidopsis seed shape by a cardioid: Efficacy of the adjustment with a scale change with factor equal to the Golden Ratio and analysis of seed shape in ethylene mutants. *Journal of Plant Physiology*, 167(5), 408–410.

[GR] Gradshteyn, I. S., & Ryzhik, I. M. (2007). *Table of Integrals, Series, and Products.* Jeffrey, A. & Zwillinger, D. (Eds.), 7th edition. New York: Academic Press.

[HW] Hall, L., & Wagon, S. (1992). Roads and wheels. *Mathematics Magazine*, 65(5), 283–301.

[Tab] `http://integral-table.com/`. Coypright own by Nikos Drakos, Computer Based Learning Unit, University of Leeds, UK and Ross Moore, Mathematics Department, Macquarie University, Sydney, Australia.

[MH] Marshall, R. N., & Harry, W. R. (1939). A cardioid directional microphone. *Journal of the Society of Motion Picture Engineers*, 33(9), 254–277.

[Cau] `http://www.mathcurve.com/surfaces/caustic/caustic.shtml`

[Mat] Mathar, R. J. (2014). Yet another table of integrals. *arXiv*:1207.5845v3 [math.CA]. Available online at `http://arxiv.org/abs/1207.5845`.

[PRÁ] Pastor, G., Romera, M., Álvarez, G., & Montoya, F. (2004). Chaotic bands in the Mandelbrot set. *Computers & Graphics*, 28(5), 779–784.

[PC] Polyanin, A. D., & Chernoutsan, A. I. (2010). *A Consise Handbook of Mathematics, Physics and Engineering Sciences.* Boca Raton, Florida: CRC Press, Taylor & Francis Group.

[RPÁ] Romera, M., Pastor, G., Álvarez, G., & Montoya, F. (2004). External arguments of Douady cauliflowers in the Mandelbrot set. *Computers & Graphics*, 28(3), 437–449.

[Sag] `http://www.sagemath.org/`

[SLL] Spiegel, M. R., Lipschutz, S., & Liu, J. (2009). *Mathematical Handbook of Formulas and Tables*, 3rd edition. New York: McGraw-Hill.

[Ste1] Stewart, J. (2010). *Calculus–Concepts and Contexts*, 4th edition, (Metric International edition). Boston, Massachusetts: Brooks/Cole Cengage Learning.

[Ste2] Stewart, J. (2012). *Single Variable Calculus*, 7th edition. Boston, Massachusetts: Brooks/Cole Cengage Learning.

[Sym] `http://symbolab.com/`

[TWH] Thomas, G. B., Jr., Weir, M. D., & Haas, J. (2010). *Thomas' Calculus*, 12th edition, (Metric edition). Boston, Massachusetts: Pearson Education, Inc.

[Wol] `http://www.wolframalpha.com/`

[Yat1] Yates, R. C. (1947). *A Handbook on Curves and Their Properties.* Ann Arbor, Michigan: J. W. Edwards.

[Yat2] Yates, R. C. (1959). The cardioid. *The Mathematics Teacher*, 52(1), 10–14.

Department of Mathematics, University College, Sungkyunkwan University, Natural Science Campus, Suwon 16149, Republic of Korea
Email address: `natanael@skku.edu`

Department of Economics, School of Humanities and Social Sciences, Nazarbayev University, Astana 010000, Kazakhstan
Email address: `binur.yermukanova@nu.edu.kz`

Mathematical Problem-Solving via Wallas' Four Stages of Creativity: Implications for the Undergraduate Classroom

Milos Savic[1]
University of Oklahoma

Abstract: The central theme in this article is that certain problem-solving frameworks (e.g., Polya, 1957; Carlson & Bloom, 2005) can be viewed within Wallas' four stages of mathematical creativity. The author attempts to justify the previous claim by breaking down each of Wallas' four components (preparation, incubation, illumination, verification) using both mathematical creativity and problem-solving/proving literature. Since creativity seems to be important in mathematics at the undergraduate level (Schumacher & Siegel, 2015), the author then outlines three observations about the lack of fostering mathematical creativity in the classroom. Finally, conclusions and future research are discussed, with emphasis on using technological advances such as Livescribe™ pens and neuroscience equipment to further reveal the mathematical creative process.

Keywords: mathematical creativity, problem solving, proving, fostering creativity, incubation, creative process

[1] savic@ou.edu

The Mathematics Enthusiast, **ISSN 1551-3440, vol. 13, no.3**, pp. 255 – 278

Introduction

Problem solving is a process that is declared important for mathematics and mathematics education. Schoenfeld (1992), after citing national studies by the National Council of Teachers of Mathematics and the National Research Council, stated that "there is general acceptance of the idea that the primary goal of mathematics instruction should be to have students become competent problem solvers" (p. 3). While mathematics education researchers continue to investigate problem solving to understand mechanisms that a solver goes through, many components of this problem-solving process built upon seminal work of Polya (1957) that provided four stages: (i) understanding the problem, (ii) developing a plan, (iii) carrying out the plan, and (iv) looking back. Polya's four-stage framework influenced other problem-solving frameworks, including Carlson and Bloom's (2005) *Multidimensional Problem-solving Framework*. However, the term "problem solving" has been defined in many ways, to the point where Chamberlin (2008) stated: "There is rarely an agreed upon definition of mathematical problem solving and reaching consensus on a conceptual definition would provide direction to subsequent research and curricular decisions" (p. 1).

Similarly, the term "mathematical creativity" has been defined in many ways, to the point where Mann (2006) stated: "An examination of the research that has attempted to define mathematical creativity found that the lack of an accepted definition for mathematical creativity has hindered research efforts" (p. 238). However, Mann (2006) also claimed that not investigating mathematical creativity to enhance students' efforts could "drive the creatively talented underground or, worse yet, cause them to give up the study of mathematics altogether" (p. 239). Mathematical creativity is either a process or a product (depending on the definition) that is declared important for mathematics and mathematics education. The Committee on

Undergraduate Programs in Mathematics (Schumacher & Siegel, 2015) stated that, "A successful major offers a program of courses to gradually and intentionally leads students from basic to advanced levels of critical and analytical thinking, while encouraging creativity and excitement about mathematics" (p. 9). According to Liljedahl (2009), "it is through mathematical creativity that we see the essence of what it means to 'do' and learn mathematics." (p. 239). While mathematical creativity has been researched (e.g., Sriraman, 2004), many components of the creative process come from the psychologist Wallas (1926). Wallas stated that there are four stages of creativity: i) preparation (thoroughly understanding a problem), ii) incubation (when the mind goes about solving a problem subconsciously and/or automatically), iii) illumination (internally generating an idea after the incubation process, sometimes known as the AHA! experience), and iv) verification (determining whether that idea is correct). Mathematicians (Hadamard, 1945; Poincaré, 1946) have stated that they have experienced similar stages in their mathematical process, and primary and secondary school mathematics education researchers have used Wallas' stages to explore problem solving (e.g., Prusak, 2015).

As discussed in the previous two paragraphs, there seems to be a connection between both problem solving and mathematical creativity, which evokes the following research question: are stages of "problem solving" and "mathematical creativity" equal sets, or is one a subset of another? In this article, the stance is that the stages of problem solving are a subset of mathematical creativity, applying Wallas' four-stage process as a basis for discussion of problem solving. Utilizing the psychodynamic lens, the consideration of mathematical creativity is more in the process of problem solving (e.g., Guilford, 1967; Pelczer & Rodriguez, 2011) and less in the product created by said process (e.g., Runco & Jaeger, 2012). A review of some of the mathematics education literature, organized into Wallas' four stages, is discussed at the beginning of this article. An emphasis is added on the proving process, seen as a subset of problem solving (Furinghetti & Morselli, 2009; Weber, 2005), since most mathematicians, graduate students, and upper-level undergraduates students employ problem solving in their proving. A discussion of teaching observations incorporating the four stages in the undergraduate mathematics classroom, with possible future research including projects using new technology, conclude the article.

Wallas' Four-stage Creative Process

The Wallas model is categorized as a *psychodynamic* approach in Sternberg's (2000) six approaches of creativity. According to Freiman and Sriraman (2007), "the psychodynamic approach to studying creativity is based on the idea that creativity arises from the tension between conscious reality and unconscious drives" (p. 24). There is a dynamic interplay between consciousness and subconscious/nonconscious, hence the term "psychodynamic."

Wallas' psychodynamic creative process has been previously verified by researchers. By interviewing mathematicians, Sriraman (2004) found that a similar four-stage creative process often occurred. However, it seems as though there are non-anecdotal difficulties with measuring the psychodynamic approach (Liljedahl, 2004), partly due to the difficulty of capturing subconscious or nonconscious. The 21st century may bring technological innovations to research methodology (e.g., neuroscience methods, tablet/Livescribe™ pens) to further probe this approach.

Other researchers have demonstrated that impasses are important in psychodynamic mathematical creativity, since impasses might generate a break in problem-solving and allow the subconscious to play. Mason, Burton, and Stacey (1982) suggest, "there is nothing wrong with being unable to make progress on a question, and there is a tremendous value in tussling with it, rephrasing it, distilling it, mulling it over, and modifying it in various ways" (p. 142). Ervynck (1991) stated that:

> What is essential in the individual is a state of mind prepared for mental activity that relates previously unrelated concepts. [Mathematical creativity] is often observed to occur after a period of intense activity involving a heightened state of consciousness of the context and all the constituent parts. And yet it is more likely to bear fruit when the mind is subsequently relaxed and able, subconsciously, to relate the ideas in a manner which benefits from quiet, unforced, contemplation. (p. 44)

Impasses seem to be an important aspect of mathematics; if any mathematician did not experience an impasse, then Fermat's Last Theorem may have been proved years ago. But one must have "intense activity" in order for an impasse, and subsequent subconscious work, to occur. What problem-solving aspects do that "intense activity", or the preparation stage, encompass?

Preparation

The *Preparation* stage of Wallas' model "focuses the individual's mind on the problem and explores the problem's dimensions," (Baker & Czarnocha, 2015, p. 4) Haylock (1987) described preparation as the stage where "the problem is investigated thoroughly and consciously, and familiarity with all its aspects is obtained" (p. 63). This may be the most important of the four stages: without it, no problem solving occurs, and with little preparation, there may be no way for the brain to take advantage of the other three stages. Poincaré (1958) believed that the "preparation stage" along with incorrect proving attempts on proofs is more useful than one usually thinks, believing it sets the unconscious mind at work.

Preparation can be observed to exist in many different forms in different frameworks. Developing a framework influenced by the work of Pólya (1957), Carlson and Bloom (2005) stated that their first stage in problem solving is "orientation," which they defined as "initially engag[ing] in intense efforts to make sense of the information in the problem" (p. 68). Both "understanding a problem" and "making sense of the information" are important in the preparation stage, but planning, executing, and checking (the other three stages in Carlson and Bloom's (2005) framework) may also be involved in the preparation stage defined by Wallas. For example, after orientation, one may immediately attempt (plan and execute) a solution to a problem, verify that the solution is incorrect, and be back at the planning stage, thus exploring more of the problem. In fact, if an individual perceives that a solution is correct after the first attempt with relative ease, this problem may not have been a problem in the sense of Schoenfeld (1985), but rather an exercise for the solver. Hence, the preparation stage may discern whether a problem was an exercise or a true "problem" for a person.

However, there are some instances in which mathematicians attempt a problem for hours. This requires many different iterations of the planning-executing-checking cycle, which may convert a problem into an exercise without the need for a "break." Silveira (1972) found in an empirical psychological study that "problem solvers performed better with longer preparation *and incubation periods* [emphasis added]" (Sio & Ormerod, 2009, p. 96). When one has exhausted the exploration of the problem's dimensions, one may tend to take a (perhaps a well-deserved) break.

Incubation

That break, according to Wallas' four stages, is called *incubation.* Incubation is defined as when "the problem is internalized into the unconscious mind and nothing appears externally to be happening" (Baker & Czarnocha, 2015, p. 4). Incubation has also been described as "a gradual and continuous unconscious process . . . during a break in the attentive activity toward a problem" (Segal, 2004, p. 141).

What happens in the mind during the incubation period? Neuroscientists have been investigating what the brain does physically (blood oxygenation levels via fMRI) or electrically (neural impulses via EEG) during the incubation period. For example, there have been different fMRI studies on incubation in problem solving (e.g., Binder, et al., 1999; De Luca, Beckmann, De Stefano, Matthews, & Smith, 2006). Researchers found that during incubation "the brain contains highly organized, spontaneous patterns of functional activity at rest" (Buckner & Vincent, 2007, p. 2), which provides evidence for incubation being a somewhat successful problem-solving action.

Incubation has its roots in the psychology literature. For example, Segal (2004) conducted an experiment to investigate whether breaks (incubation periods) from problem-solving could be helpful, and what difficulty of cognitive tasks might be introduced during an incubation period. Segal concluded that "a break would improve the performance of insight problem-solving [the psychodynamic approach], but that the duration of the break would not influence performance" (p. 147). On cognitive tasks, Segal stated, "less demanding activity during the break serves as a weaker diversion" more than higher-demanding tasks. The hypothesis about higher cognitive-demand tasks during the incubation period has been supported in other studies (e.g., Kaplan, 1990; Patrick, 1986). However, many of the tasks that have been used to generate incubation (and subsequent insight) involved laboratory problems of time lengths from one to 60 minutes (Sio & Ormerod, 2009). In contrast, many mathematicians, and perhaps even undergraduate or graduate students, incubate in proving or problem-solving for more than 60 minutes, and perhaps also in other settings (home, park, theater, etc., (Savic, 2015)). Therefore, while the field of psychology has much to offer in terms of vocabulary and motivation, experiments investigating mathematical creativity may need to be modified in order to maximally capture problem solving in a more natural setting.

Savic (2015) attempted to address the issue of exploring mathematicians' problem-solving process in a more natural environment by equipping nine mathematicians with Livescribe™ pens, capable of capturing synchronized writing and audio and providing them with challenging mathematical tasks. The resulting data was uploaded to a computer, and date and time stamps were associated with each synchronized proving session. Six of the nine mathematicians experienced some impasses in their proving, and all six engaged in an incubation period of various lengths. Exit interviews with the mathematicians exposed that some incubation periods yielded illuminations or AHA! moments, and that mathematicians had established procedures for where and how they would engage in incubation. For example, one mathematician, Dr. G, talked about taking a walk, and stated that this was common when he encountered an impasse. He stated:

> When [mathematicians] are working on something, we are usually scribbling down on paper. When you go take a break, . . . you are thinking about it in your head without any visual aids . . . [walking around] forces me to think about it from a different point of view, and try different ways of thinking about it, often [from] global, structural points of view. (Savic, 2015, p. 75)

The incubation stage tends to be the most difficult to acknowledge and investigate. Many studies have investigated incubation via self-reported retrospective evidence by interviewing mathematicians (Sriraman, 2004) or students (Garii, 2002). Another difficulty of investigating incubation is that it is usually coupled with illumination or insight. When a problem solver has a sudden insight or AHA! moment, it is usually after a period of incubation. Hence, acknowledgment of a period of incubation must occur *after* an insight. Aiken (1973, p. 409) added a caveat of "success" to the definition of incubation and insight by stating, "if the prolonged unconscious work of the second stage is successful, a third stage occurs-illumination or insight into a solution." I respectfully disagree with Aiken on one aspect; insight does not have to be "correct" for the acknowledgment of incubation. However, a question may be posed: What defines the "end" of incubation?

Illumination

The idea that arises from an incubation period is called an *illumination*. Sometimes called insight, illumination is defined by Baker and Czarnocha (2015) as "where the creative idea bursts forth from its preconscious processing into conscious awareness" (p. 5). Leikin (2014) stated that "insight exists when a person acts adequately in a new situation, and as such, insight is closely related to creative ability" (p. 249). Liljedahl (2013) stated that "illumination is THE aspect of the process that sets creativity, discovery, and invention apart from the more ordinary, and more common, processes of solving a problem—it is the marker that something remarkable has taken place" (p. 255). Therefore, one may associate many "product" definitions of mathematical creativity (e.g., "originality or effectiveness" (Runco & Jaeger, 2012, p. 92)) with insights, since the insight is the (unverified) idea of a solution for a problem or theorem, and hence a mathematical product.

Mathematicians have described *insight* as "seeing a connection," "the light switches," and "having a feel for how things connect together" (Burton, 1999b, p. 28). How does one obtain this illumination? Many believe it is the subconscious "at play" when the person is in a state of incubation. It is sometimes referred to as the "AHA Moment" (Liljedahl, 2004), due to its unexpected nature, coupled with a sense of euphoria or positive emotion (Burton, 1999a). Liljedahl (2013, p. 264) stated that the affective component of illumination is what sets it "apart from other mathematical experiences." Also, the person must have some conviction (Poincaré, 1946) that the illumination has value towards the solution or proof, or else there would be no elation from an AHA moment.

The moment of insight has been described as an aspect of Koestler's (1964) bisociation theory: "the spontaneous leap of insight...which connects previously unconnected matrices of experience [frames of reference] and makes us experience reality on several planes at ones" (p. 45; cited in Baker & Czarnocha, 2015). Sriraman (2004, p. 30) conjectured that "the mind throws out fragments (ideas) which are products of past experience." Somehow the mind, when problem solving, subconsciously keeps piecing together information, perhaps even digging into long-term memory to create a solution or proof. However, that AHA moment for a solution or proof may be invalid.

Verification

The *verification* stage is needed to acknowledge that the illumination is valid. Verification is defined by Baker and Czarnocha (2015) as "where the idea is consciously verified, elaborated, and then applied" (p. 5). It is the stage in which an illumination is confirmed or refuted, including many of the little details that may not have been fully checked in one's mind. In the Carlson and Bloom (2005) multidimensional problem-solving framework, the two phases "executing" and "checking" are both located in the verification stage, since the executing stage is when writing occurs in order to expand on ideas created in the planning stage, and the checking stage is when one examines for errors what s/he has written.

Liljedahl (2004) cautioned that the verification stage is not only about validity of the solution: "it is also a method by which the solver re-engages with the problem at the level of details" (p. 16). In fact, one could view verification as metacognitive, where one may also be looking for uses of the problem-solving or proving technique for other problems, or posing questions that may develop from the solution. Knuth (2002) stated that "much is to be gained by making the solution or an aspect of the problem a starting point for mathematical exploration – exploration that lies at the heart of mathematical practice" (p. 130).

In the proof literature, verification may be discussed as proof validation, defined as "the process an individual carries out to determine whether a proof is correct and actually proves the particular theorem it claims to prove" (Selden & Selden, 1995, p. 127). Proof validation is a subset of proof comprehension, which considers "understanding the content of the proof and learning from it" (Mejia-Ramos & Weber, 2011, p. 331). To validate, one must understand the content of the proof. While proof validation has been examined in the literature (e.g., Selden & Selden, 2003), proof self-validation, which I consider the verification stage of the proving process, has rarely been examined in detail.

Verification is the stage that may differentiate creativity in mathematics from other disciplines. According to Sadler-Smith (2015), Wallas had a three-stage creativity process until he used the writings of Poincaré to add the fourth stage, stating that "incubation supplied a starting point for further work in the verification stage" (p. 344). Poincaré himself stated that "It usually happens that it [the illumination] does not deceive him [the mathematician], but it also sometimes happens, as I have said, that it [the illumination] does not stand the test of verification" (Sadler-Smith, 2015, p. 344). If one does not have a success in the verification stage, the cycle comes back to either preparation ("What am I missing?") or incubation ("Maybe I need another break?").

Mathematical Creativity Teaching Observations

Wallas' four stages may be happening in undergraduate mathematics; however, there seems to be little evidence for its explicit discussion in the classroom. Below, I outline three observations that I believe may be occurring in university courses that may assist with discussing mathematical creativity in the classroom. These observations come from experience teaching undergraduate classes and the shortage of research literature about mathematical creativity and Wallas' four stages at the undergraduate level.

Observation 1: A majority of tasks posed may not allow for the four stages to occur.

Schoenfeld (1989) surveyed 206 high school students on the length of time it took to solve a typical homework problem, and the average was just under 2 minutes, and not one of the 206 students "allotted more than 5 minutes" (p. 345). Coincidentally, in Lithner's (2004) calculus textbook study, he claimed that over 70% of the problems can be solved using examples in the text. According to Selden and Selden (2013), in proof-based courses, several tasks are "Type 1," where the "proofs…can depend on a previous result in the notes" (p. 320), as opposed to "Type 2," defined as "require formulating and proving a lemma not in the notes, but one that is relatively easy to notice, formulate, and prove" (p. 320) and "Type 3," which is hard to notice the lemma needed. This may convince a student that one needs to arrive at the solution or proof immediately.

Tasks of this type may not reflect how mathematics is generally practiced (Burton, 1999b). However, there are articles that have discussed tasks that allow for the potential of creativity (Leikin, 2014). Zazkis and Holton (2009) posed tasks from topics such as graph theory and number theory. One of their tasks is stated: "Prove that $n^5 - n$ is divisible by 3" (p. 348). Silver (1997) stated that perhaps students might need "complex, ill-structured problems" (p. 77) in order to engage in the mathematical creative process. Regardless of the topic of task, students may need some tasks that allow them to incubate and illuminate, since they may have situations later in life that may push them to participate in the creative process (Tomasco, 2010). The difficulty with assigning these tasks is that instructors then open up the classroom for a variety of problem-solving strategies, some of which the instructor may not be prepared for, or may be completely incorrect. Therefore, the next observation may help alleviate this difficulty.

Observation 2: Many students do not have classroom environments where impasses and productive failure is encouraged by their instructors.

While tasks play a part in fostering an environment, the instructor's actions, both in the classroom and in feedback, play another crucial role. Actions in the classroom can include think-pair-share periods, defined as "a multi-mode discussion cycle in which students have time to *think* individually, talk with each other in *pairs,* and finally *share* responses with the larger group" (McTighe & Lyman, 1988, p. 19). Think-pair-share may encourage students that are shy with the large group to see that others are experiencing impasses, and hopefully exchange ideas for overcoming such impasses.

When presented with a non-traditional solution from a student, an instructor has many choices to make, including how to address the solution presented, what to do with some students who did not understand the solution, and so forth. However, the instructor may have preconceived solutions that s/he wanted to emphasize. This scenario seems all too common in the mathematics classroom, but how the scenario is settled has ramifications for how the social and sociomathematical norms (Yackel & Cobb, 1996) are fostered. Hershkovitz, Pelled, and Littler (2009) stated that "a teacher who has some pre-determined answer expectations might easily suppress creative initiatives by not accepting or ignoring children's ideas" (p. 265). This suppression may have long-lasting effects: Students that might have shared their ideas may be "scared" or "afraid," thus silencing a majority of students that may have creative solutions to certain problems. Mistakes and productive failure can be incredibly important in education (Burger & Starbird, 2012), and can be used in the classroom as "springboards to inquiry," allowing students to "go beyond diagnosis and remediation" (Borasi, 1994, p. 167). Instead, to encourage mathematical creativity, instructors "should encourage good ideas even (and, in fact, especially) when a student suggests an unexpected answer or when the answers are inaccurate" (Hershkovitz, Peled, & Littler, 2009, p. 265).

Observation 3: Many students have not developed actions to cope with impasses, that is, students may not have found their "incubation" or "preparation" actions.

The author (Savic, 2012; Savic, 2015) studied both mathematicians and graduate students proving theorems, focusing on their incubation periods. A majority of the mathematicians had certain activities that they engaged in specifically for incubation, including watching TV, sleeping, or going on a walk in a certain park. In contrast, none of the five graduate students had an activity to implement when incubating. When interviewed, and pressed about incubation, they spoke of taking breaks in a general manner. In combination with Schoenfeld's (1989) study about high school students and time spent on homework, perhaps many undergraduate/graduate students may not have experienced many situations where incubation needed to take place. Therefore, encouraging students to have activities or places of incubation may help their problem-solving processes.

The "preparation" stage may be lacking in some students. In proving, preparation rarely results in the key idea of a proof; often "unpacking" (in the sense of Selden and Selden, 1995) definitions may yield a valid proof of a Type 1 theorem (Selden and Selden, 2013). However, some students may not engage in the proving process at all by avoiding certain theorems, or if they do, their lack of persistence may be a block for successful proving. Our research group created a rubric, named the Creativity-in-progress Rubric (CPR) on Proving (Savic, Karakok, Tang, El Turkey, & Naccarato, in press; Karakok, Savic, Tang, & El Turkey, in press), to consider actions in the preparation stage of the proving process that may yield the potential to be creative.

There are two main categories of the CPR on Proving: *making connections* and *taking risks*. *Making connections* is defined as an ability to connect the proving task with definitions, theorems, multiple representations, and examples from the current course that a student is in, and possible prior experiences from previous courses. The sub-categories involve making connections: between definitions/theorems, between representations, and between examples. *Taking risks* is defined as an ability to actively attempt a proof, demonstrate flexibility in using multiple approaches or techniques, posing questions about reasoning within the attempts, and evaluating those attempts. The sub-categories involve: tools and tricks, flexibility, posing questions, and evaluation of the proof attempt. Every sub-category has three different levels: beginning, developing, and advancing. Each level has a small description to help either the instructor or student know what level s/he is in, and what s/he can achieve (See Appendix A for the full rubric and details).

The actions outlined in the CPR on Proving allow students to explicitly examine their own proving process and adjust accordingly. Self-evaluation of the proving process may help create persistence, and hence situate the student in the preparation stage, with a better chance for incubation (conscious and sub-conscious thoughts) to have more effect. Incubation cannot be only recommended, taught, or self-forced. For students to allow incubation (and subsequently insight) to occur, there must be a certain amount of preparation. This preparation cannot be measured by time, but perhaps by perseverance on the solution (almost to exhaustion) coupled with a curiosity. I contend that mathematical curiosity (Knuth, 2002) may be, ultimately, most important for a student, both in the short-term with finding a solution, and in the long-term with his/her penchant to attack more problems with less difficulty or resistance.

Conclusions/Future Research

In conclusion, Wallas' (1926) four stages of the creative process (preparation, incubation, illumination, and verification), brought into the mathematical world by Hadamard (1945), seem to permeate the literature of both psychology and mathematics education. Preparation is the most important stage of the four: Without preparation, there is no engagement in the other three stages. Investing many of one's mental resources may not lead one to a satisfactory solution to a problem; therefore, one may recognize this as an impasse and take a "break." Incubation allows conscious and subconscious ideas to merge, signaling the brain when a spontaneous solution has been brought to consciousness. This AHA! moment or insight is the illumination stage, where a creative idea is found and used. The validation stage identifies if the insight is correct; if not, either the person cycles back to the preparation stage, or goes into another incubation stage.

Teaching the psychodynamic approach of mathematical creativity may be difficult. An instructor must consider merging three different dimensions of the classroom in harmony for a student to participate in the four stages:

1) Some tasks must allow incubation to occur.
2) Pedagogical moves may need to permit impasses and productive failure to occur in the classroom.
3) Instructors could assist students in finding personal actions that foster the preparation and incubation stages.

There are three future research directions of the creativity in undergraduate mathematics research group that created the CPR on Proving. First, investigation in the implementation of the CPR on Proving in the classroom is a priority. How are instructors and students utilizing the rubric, and what teaching materials can accompany the rubric? Secondly, the group would like to explore the social facet of creativity in proving with both mathematicians and undergraduate students. Is there social activity that fosters creativity in collaboration? Finally, focusing on Wallas' four stages, the author and another colleague are collaborating with the neuroscience division at our university to investigate brain activity in the proving process, including what insights may occur, whether insight is correlated with key ideas of a proof (in the sense of Raman, 2003), and whether one can predict, along with other factors, when a student may have an insight in his/her proving process. With the invention of neuroeducational theory (Anderson, 2014), coupled with new naturalistic data collection techniques such as LiveScribe™ pens (Savic, 2015), the future is bright for research in mathematical creativity *and* problem solving.

References

Aiken, L. (1973). Ability and creativity in mathematics. *Review of Educational Research*, 405-432.

Anderson, O. R. (2014). Progress in application of the neurosciences to an understanding of human learning: The challenge of finding a middle-ground neuroeducational theory. *International Journal of Science and Mathematics Education, 12*, 475-492.

Baker, W., & Czarnocha, B. (2015). AHA! moment, bisociation and simultaneity of attention. *Congress of European Research in Mathematics Education – 9 Proceedings.* Prague.

Binder, J. R., Frost, J. A., Hammeke, T. A., Bellgowan, P. S., Rao, S. M., & Cox, R. W. (1999). Conceptual processing during the conscious resting state: A functional MRI study. *Journal of Cognitive Neuroscience*, 80-93.

Borasi, R. (1994). Capitalizing on errors as" springboards for inquiry": A teaching experiment. *Journal for Research in Mathematics Education*, 166-208.

Burger, E. B., & Starbird, M. (2012). *The 5 elements of effective thinking.* Princeton, NJ: Princeton University Press.

Burton, L. (1999). The practices of mathematicians: What do they tell us about coming to know mathematics? *Educational Studies in Mathematics, 37*, 121-143.

Burton, L. (1999). Why is intuition so important to mathematicians but missing from mathematics education? *For the Learning of Mathematics, 19*(3), 27-32.

Byers, W. (2007). *How mathematicians think: Using ambiguity, contradiction, and paradox to create mathematics.* Princeton, NJ: Princeton University Press.

Carlson, M., & Bloom, I. (2005). The cyclic nature of problem solving: An emergent problem-solving framework. *Educational Studies in Mathematics, 58*, 45-75.

Chamberlin, S. A. (2008). What is problem solving in the mathematics classroom? *Philosophy of Mathematics Education, 23.* Retrieved June 8, 2012, from http://people.exeter.ac.uk/PErnest/pome23/Chamberlin%20What%20is%20Math%20Prob%20Solving.doc.

Craft, A. (2001). *An analysis of research and literature on creativity in education.* Report prepared for the Qualifications and Curriculum Authority.

De Luca, M., Beckmann, C. F., De Stefano, N., Matthews, P. M., & Smith, S. M. (2006). fMRI resting state networks define distinct modes of long-distance. *NeuroImage, 29*, 1359-1367.

Ervynck, G. (1991). Mathematical creativity. In D. Tall, *Advanced Mathematical Thinking* (pp. 42-52). New York, NY, USA: Kluwer Academic Publishers.

Freiman, V., & Sriraman, B. (2007). Does mathematics gifted education need a working philosophy of creativity? *Mediterranean Journal for Research in Mathematics Education, 6*, 23-46.

Furinghetti, F., & Morselli, F. (2009). Every unsuccessful problem solver is unsuccessful in his or her own way: Affective and cognitive factors in proving. *Educational Studies of Mathematics, 70*, 71-90.

Garii, B. (2002). *That "Aha" experience: Meta-cognition and student understanding of learning and knowledge.* Retrieved May 15, 2012, from http://eric.ed.gov/PDFS/ED464127.pdf

Guilford, J. P. (1967). *The nature of human intelligence.* New York, NY, U.S.A.: McGraw-Hill.

Hadamard, J. (1945). *The mathematician's mind.* Princeton: Princeton University Press.

Haylock, D. (1987). A framework for assessing mathematical creativity in school children. *Educational Studies in Mathematics, 18*(1), 59-74.

Hershkovitz, S., Peled, I., & Littler, G. (2009). Mathematical creativity and giftedness in elementary school: Task and teacher promoting creativity for all. In R. Leikin, A. Berman, & B. Koichu, *Creativity in Mathematics and the Education of Gifted Students* (pp. 255-269). Sense Publishers.

Kaplan, C. (1990). *Hatching a theory of incubation: Does putting a problem aside really help? If so, why?* Carnegie Mellon University.

Karakok, G., Savic, M., Tang, G., & El Turkey, H. (in press). Mathematicians' views on undergraduate student creativity. *Congress of European Research in Mathematics Education – 9 Proceedings.* Prague.

Knuth, E. (2002). Fostering mathematical curiosity. *The Mathematics Teacher, 95*(2), 126-130.

Koestler, A. (1964). *The Act of Creation.* London: Hutchinson & Co (publishers), LTD.

Leikin, R. (2014). Challenging mathematics with multiple solution tasks and mathematical investigations in geometry. In Y. Li, E. A. Silver, & S. Li, *Transforming mathematics instruction: Multiple approaches and practices* (pp. 59-80). Dordrecht, the Netherlands: Springer.

Leikin, R. (2014). Giftedness and high ability in mathematics. In S. Lerman, *Encyclopedia of mathematics education* (pp. 247-251). Springer Netherlands.

Liljedahl, P. (2004). *The AHA! experience: Mathematical contents, pedagogical implications.* (Doctoral Dissertation). Vancouver: Simon Frasier University.

Liljedahl, P. (2009). In the Words of the Creators. In R. Leikin, A. Berman, & B. Koichu, *Mathematical Creativity and the Education of Gifted Children* (pp. 51-70). Rotterdam: Sense Publishers.

Liljedahl, P. (2013). Illumination: an affective experience? *ZDM, 45*(2), 253-265.

Lithner, J. (2004). Mathematical reasoning in calculus textbook exercises. *The Journal of Mathematical Behavior, 23*, 405-427.

Mann, E. (2006). Creativity: The essence of mathematics. *Journal for the Education of the Gifted, 30*(2), 236-260.

McTighe, J., & Lyman, F. T. (1988). Cueing thinking in the classroom: The promise of theory-embedded tools. *Educational Leadership, 45*(7), 18-24.

Mejia-Ramos, J. P., & Weber, K. (2011). Why and how mathematicians read proofs: An exploratory study. *Educational Studies in Mathematics, 76*(3), 329-344.

Patrick, A. S. (1986). The role of ability in creative "incubation". *Personality and Individual Differences, 7*(2), 169-174.

Pelczer, I., & Rodriguez, F. G. (2011). Creativity assessment in school settings through problem posing tasks. *The Montana Mathematics Enthusiast, 8*, 383-398.

Poincaré, H. (1946). *The foundations of science.* Lancaster, PA: The Science Press.

Polya, G. (1957). *How to solve it: A new aspect of mathematical method.* Garden City, NJ: Doubleday.

Prusak, A. (2015). Nurturing students' creativity through telling mathematical stories. *Proceedings of the 9th International MCG Conference*, (pp. 16-21). Sinaia, Romania.

Raman, M. (2003). Key Ideas: What are they and how can they help us understand how people view proof? *Educational Studies in Mathematics, 52*, 319-325.

Runco, M. A., & Jaeger, G. G. (2012). The standard definition of creativity. *Creativity Research Journal, 24*(1), 92-96.

Sadler-Smith, E. (2015). Wallas' Four-Stage Model of the Creative Process: More Than Meets the Eye? *Creativity Research Journal, 27*(4), 342-352.

Savic, M. (2012). *Proof and Proving: Logic, Impasses, and the Relationship to Problem Solving.* (Doctoral Dissertation): New Mexico State University.

Savic, M. (2015). The incubation effect: How mathematicians recover from proving impasses. *Journal of Mathematical Behavior, 39*, 67-78.

Savic, M., Karakok, G., Tang, G., El Turkey, H., & Naccaratto, E. (in press). Formative Assessment of Creativity in Undergraduate Mathematics: Using a Creativity-in-Progress Rubric (CPR) on Proving. In R. Leikin, & B. Sriraman, *Creativity and Giftedness: Interdisciplinary Perspectives from Mathematics and Beyond.* Springer.

Schoenfeld, A. (1989). Explorations of students' mathematical beliefs and behavior. *Journal for Research in Mathematics Education, 20*(4), 338-355.

Schoenfeld, A. H. (1985). *Mathematical problem solving.* Orlando, FL: Academic Press.

Schoenfeld, A. H. (1992). Learning to think mathematically: Problem solving, metacognition, and sense-making in mathematics. In D. Grouws (Ed.), *Handbook for research on mathematics teaching and learning* (pp. 334-370). New York, NY: Macmillian.

Schumacher, C. S., & Siegel, M. J. (2015). *2015 CUPM Curriculum Guide to Majors in the Mathematical Sciences.* Washington, DC: Mathematical Association of America.

Segal, E. (2004). Incubation in insight problem solving. *Creativity Research Journal, 16*, 141-148.

Selden, A., & Selden, J. (2003). Validations of proofs considered as texts: Can undergraduates tell whether an argument proves a theorem? *Journal for Research in Mathematics Education, 34*, 4-36.

Selden, A., & Selden, J. (2013). Proof and problem solving at the university level. *The Mathematics Enthusiast, 10*(1&2), 303-334.

Selden, J., & Selden, A. (1995). Unpacking the logic of mathematical statements. *Educational Studies in Mathematics, 29*, 123-151.

Silveira, J. (1972). *Incubation: The effect of interruption timing and length on problem solution and quality of problem processing.* Unpublished doctoral dissertation, University of Oregon.

Silver, E. (1997). Fostering creativity through instruction rich in mathematical problem solving and posing. *ZDM Mathematical Education, 3*, 75-80.

Sio, U. N., & Ormerod, T. C. (2009). Does incubation enhance problem solving? A meta-analytic review. *Psychological Bulletin, 35*, 94-120.

Sriraman, B. (2004). The characteristics of mathematical creativity. *The Mathematics Educator, 14*, 19-34.

Sternberg, R. J. (2000). *Handbook of Creativity.* Cambridge University Press.

Tomasco, S. (2010, May 18). *IBM 2010 Global CEO Study: Creativity Selected as Most Crucial Factor for Future Success*. Retrieved March 19, 2016, from IBM: http://www-03.ibm.com/press/us/en/pressrelease/31670.wss

Wallas, G. (1926). *The art of thought.* New York: Harcourt Brace.

Weber, K. (2005). Problem-solving, proving, and learning: The relationship between problem-solving processes and learning opportunities in the activity of proof construction. *Journal of Mathematical Behavior, 24*, 351-360.

Yackel, E., & Cobb, P. (1996). Sociomathematical norms, argumentation, and autonomy in mathematics. *Journal for Research in Mathematics Education*, 458-477.

Zazkis, R., & Holton, D. (2009). Snapshots of creativity in undergraduate mathematics education. In R. Leikin, A. Berman, & B. Koichu, *Creativity in mathematics and the education of gifted students* (pp. 345-365). Rotterdam, the Netherlands: Sense.

Aesthetics in School Mathematics: A Potential Model and A Possible Lesson

Hartono Tjoe[1]

The Pennsylvania State University

Abstract: Earlier studies on improving classroom practice in mathematics have suggested a closer attention to nurturing an aesthetic appreciation for mathematics in students' learning experiences. Recent evidence nonetheless reveals little indication of its presence. This article offers a potential model of the case for aesthetics in school mathematics. Central to this model is the harmonious hierarchy of necessity, existence, and uniqueness without any of which the case for aesthetics in student learning might be suboptimal, if not untenable. This article offers an example of the proposed model using a possible lesson designed to engage students aesthetically in the learning of mathematics. Pedagogical implications are discussed to reflect and revisit an interpretation of learning mathematics through problem solving.

Keywords: mathematical aesthetics, mathematics problem-solving, teaching and learning in mathematics

[1] hht1@psu.edu

Introduction

Earlier studies on improving classroom practice in mathematics have suggested a closer attention to nurturing an aesthetic appreciation for mathematics in students' learning experiences (Krutetskii, 1976; Papert, 1980; Silver & Metzger, 1989; Smith, 1927; Sriraman, 2009). Recent evidence nonetheless reveals little indication of its presence (Dreyfus & Eisenberg, 1986; Tjoe, 2015). We discuss in this article how current considerations of aesthetics in school mathematics, if any, might have inadvertently emphasized perfunctory precision over creative process. Given its current state, we argue how aesthetics can evolve into a compelling case in school mathematics.

We begin with a survey of the notions of mathematical aesthetics and its interpretations. We present a typical contemporary classroom episode of a first grade mathematics lesson in one- and two-digit addition. We explain how exposing students to such a lesson might overlook the opportunity to reveal and foster an aesthetic appreciation for mathematics. We then offer a potential model of the case for aesthetics in school mathematics. Central to this model is the harmonious hierarchy of necessity, existence, and uniqueness without any of which the case for aesthetics in student learning might be suboptimal, if not untenable. We exemplify our model with a possible lesson designed to engage students aesthetically in the learning of mathematics. Pedagogical implications are discussed to reflect and revisit an interpretation of learning mathematics through problem solving.

Mathematical Aesthetics

Aesthetics has been one of the driving forces behind the activities that gave life to the advancements in mathematics as a discipline (Davis & Hersh, 1981). Its subtlety creates guidelines that many research mathematicians follow as one of the foremost principles in their professions. It is in the search of mathematical beauty that research mathematicians often seek approvals that lead to the crowning achievement in their mathematical experience (Hardy, 1940).

Sinclair (2004) analyzes the role of aesthetic values from several conceptual insights. She draws examples from existing empirical findings such as those by Dreyfus and Eisenberg (1986) and Silver and Metzger (1989). In one of her interpretations of their work, she suggests that "mathematicians' aesthetic choices might be at least partially learned from their community as they interact with other mathematicians and seek their approval" (Sinclair, 2004, p. 276). Furthermore, she indicates that mathematical beauty is only feasible in the process "when young mathematicians are having to join the community of professional mathematicians—and when aesthetic considerations are recognized (unlike at high school and undergraduate levels)" (p. 276).

Related to Sinclair's (2004) interpretations of mathematical aesthetics, Karp (2008) conducts a comparative study on the aesthetic aspect of mathematical problem solving. Karp's comparative study involved middle and high school mathematics teachers from the U.S. and Russia. In his study, teachers are asked to provide examples and explanations of "beautiful" mathematics problems and approaches in solving those problems. Karp's (2008) findings confirm that the curricular system of education has a tremendous impact on students' aesthetic preference in mathematics problem solving. Each group of teachers shows different perspectives on what count as mathematical "beauty." In particular, these differences stand out from their selections of mathematics topics. American teachers put extra weight on mathematics topics as prescribed by the American curriculum, which is typically associated with real-life situations and applications. Russian teachers do likewise as recommended by Russian curriculum with its traditionally heavy emphasis on algebra, number theory, and geometry. Evidently, these Russian problems tend to require longer approaches and are more algebraically demanding than their American counterparts. In their explanations, American teachers describe "usefulness in the teaching process," "useful[ness] in practical life or comes the real world," "non-standard and cannot be solved using ordinary methods that are regularly discussed in school," "unexpectedness of the solution," "openness of the problem," and "a combination of methods and knowledge from different fields of mathematics" (Karp, 2008, p. 40). Russian teachers reveal in their choices of problems and solutions the sense of "overcoming of chaos," "non-standard nature," and "traditional fields" in their origins (p. 40). In his conclusion, Karp indicates a relative character of aesthetic preference in mathematics problem solving.

Apart from the curricular system of education, the context of cultural differences has also been observed in understanding mathematical aesthetics through classroom learning or professional experiences. Tjoe (2015) examines the hypothesis as to whether there exists a gap between different problem solvers in the criteria that might be attributed to the interpretation of mathematical "beauty." Tjoe's study involves expert mathematicians at the university level and mathematically gifted students at the high school level. In his study, research mathematicians and mathematically gifted students are asked to choose their most preferred approach as they are presented with a collection of many different problem solving approaches. Tjoe's (2015) findings reveal that whereas expert mathematicians associate "beautiful" approaches with their simplicity and originality in the search for geometric reasoning or explanation, mathematically gifted students strive for the economic attribute of mathematical "beauty" in those approaches that involve fewer steps and shorter solving time. When both groups' divergent choices of preferred approaches as well as their dissimilar interpretations of mathematical "beauty" are made known to each other, it is evident that these opposing views are not construed from their mathematical content knowledge, but rather from their appeals of "beauty" based on their mathematics experiences (Tjoe, 2015). Tjoe suggests that "there appeared to be a profound lacuna in the understanding of mathematical aesthetics that might inadvertently subdivide the state of mathematically gifted into two groups: one group of professional research mathematicians and another group of those whose affects might be waiting to be nurtured" (Tjoe, 2015, p. 173). Given mathematically aesthetics is not a characteristic that problem solvers are born with, it is possible that one must learn to instill such habit in their everyday life as professional mathematicians. This possibility might further indicate that mathematical aesthetics is a socially constructed concept, or that mathematical aesthetics has found its decline in its inclusion in the teaching and learning of mathematics at the elementary and secondary levels.

A Lesson in One- and Two-digit Addition

The Common Core State Standards Initiative describes perseverance in problem solving as one of the most important standards for mathematical practice. Students are encouraged to "check their answers to problems using a different method," "understand the approaches of others," and "identify correspondences between different approaches" (Common Core State Standards Initiative, 2010, p. 6). With this in mind, first grade students are expected to be proficient in varieties of ways of solving one- and two-digit addition problems (CCSS.MATH.CONTENT.1.OA.C.5-6, 1.NBT.C.4). We include in this section some of the ways that one- or two-digit addition problems (e.g., 5 + 6 and 46 + 38) are typically presented in the common core aligned mathematics textbooks at the first grade level.

The most elementary way of solving one-digit addition problems is the counting via direct modeling of objects or fingers: 1) counting all (e.g., 5 + 6 is solved by counting aloud one, two, three, four, five (five), …, six, seven, eight, nine, ten, eleven (six)), 2) counting on from the first addend (e.g., 5 + 6 is solved by counting aloud five (five), …, six, seven, eight, nine, ten, eleven (six)), and 3) counting on from the larger addend (e.g., 5 + 6 is solved by counting aloud six (six), …, seven, eight, nine, ten, eleven (five)). The next method of solving one-digit addition problems after counting via objects or fingers is to count abstractly without objects or fingers. As one moves away from relating counting to addition toward the idea of addition through reasoning, one learns to derive addition by recalling known addition facts (e.g., $5 + 6 = 11$ because knowing $5 + 5 = 10$ and $6 = 5 + 1$, one obtains $5 + 6 = 5 + 5 + 1 = 11$) or by using commutative property of addition (e.g., $5 + 6 = 11$ because $6 + 5 = 11$ and $5 + 6 = 6 + 5$ so $5 + 6 = 11$). Finally, students are to perform mental math addition via retrieval from long-term memory so that one can instantly solve $5 + 6 = 11$ on the spot.

One way to solve two-digit addition problems is to add the tens and the ones separately, and combine them (e.g., $46 + 38 = 84$ because $40 + 30 = 70$ and $6 + 8 = 14$ so $70 + 14 = 84$). Another way to solve two-digit addition problems is to add on the tens followed by adding on the ones (e.g., $46 + 38 = 84$ because $46 + 10 + 10 + 10 = 76$ and $76 + 4 + 4 = 84$). One can also decompose an addend to make the tens (e.g., $46 + 38 = 84$ because $46 + 38 = (44 + 2) + 38 = 44 + (2 + 38) = 44 + 40 = 84$) or compensate another addend to make the tens (e.g., $46 + 38 = 84$ because $46 + 38 = 46 + (40 - 2) = (46 + 40) - 2 = 86 - 2 = 84$). A culminating point of addition lesson concludes with a traditional vertical algorithm for addition that sometimes commits carrying over to rote memorization (e.g., $46 + 38 = 84$ because $46 + 38 = (40 + 6) + (30 + 8) = (40 + 30) + (6 + 8) = (40 + 30) + 14 = (40 + 30) + (10 + 4) = (40 + 30 + 10) + 4 = 80 + 4 = 84$).

These different ways of solving one- and two-digit addition problems often wear out many teachers who eventually overemphasize to their students the memory retrieval for one-digit addition problems or the vertical addition algorithm for two-digit addition problems. A popular explanation to the overemphasis of either addition method is its *convenience*. Indeed, a number of studies demonstrate that not all of these addition strategies receive equally extensive utilization outside the teaching and learning settings (Geary & Brown, 1991; Geary & Wiley, 1991; Siegler & Shrager, 1984). Attributing their strategy choice to the most economical way of arriving at the answers, more experienced problem solvers, for instance, depend heavily on reasoning (e.g., memory retrieval) as opposed to counting to solve addition problems.

Unfortunately, such a *convenient* way of teaching the most economical approach is to a certain degree one cause of an *inconvenient* way of learning to appreciate mathematical aesthetics. Without communicating to students to help them reveal the power, usefulness, and beauty of the addition methods one after and over the other, learning mathematics might become a mere tool of pedantic precision, instead of a creative journey of problem solving. In this manner, young problem solvers might mistakenly perceive the kind of mathematics activities that research mathematicians conduct in their profession life as an impetuous act of conduct, instead of an inculcative habit of mind. It is thus essential for teachers to inform their students of the fact that the vertical addition algorithm does not simply materialize from the work of research mathematicians in the form that may be found in current mathematics textbooks. Students should also be acquainted with the aesthetics principle that directly guides the process in which research mathematicians compose, frame, and identify the vertical addition algorithm, among many other algorithms, to be what may now be considered the standard algorithm for any addition problems.

A Model of Aesthetics in School Mathematics

One way to help students to understand the real process that transform the vertical addition algorithm into the standard algorithm is to engage them in a similar experience that help define and refine the criteria of a standard algorithm from the point of view of mathematical aesthetics (Silver & Metzger, 1989). In this section, we propose a possible model of the case for aesthetics in school mathematics that integrates necessity, existence, and uniqueness. We describe the accounts of this model using backward explanations.

We maintain that our end goal is to nurture the feelings of mathematical aesthetics among our students. In the context of problem solving process, mathematical aesthetics is often identified through the uniqueness in which a problem has been approached. At one point in the problem solving process, in particular, after a problem has been successfully solved, solvers are to be able to recognize that there exists a unique problem solving approach that is preferred using a certain criteria.

On the one hand, such criteria of preference depend greatly on solvers' mathematical experience and knowledge. On the other hand, solvers can only accumulate a series of meaningful mathematical experience and knowledge, and therefore prefer a unique problem solving approach, when there is more than one problem solving approach to choose from. In other words, the existence of many different problem solving approaches is a necessary condition for the uniqueness of such problem solving approach, and perhaps more crucially, the uniqueness of the criteria which help guide solvers to prefer one problem solving approach to another. If there were only one solution method to approach a problem, then there would be no other solution method to compare with, and there would not be a need to prefer one solution method to the others. In this case, it would be a difficult effort, if not a futile one, to convince students who learn only one solution method to solve the problem that such a solution method in fact entails a great deal of aesthetics values.

Although the existence of many different solution methods can be viewed as a means to facilitate aesthetics appreciations toward the most "beautiful" problem solving approach, there needs to be a more pragmatic function (in addition to an affective one) that serves to explain why students need to learn more than one solution method. It is at this point that different numerical characteristics of problems with similar surface structures can be a determining factor. If mathematics instructors can demonstrate to their students that a certain solution method would work more effectively when applied to solve problems with a particular numerical characteristic, while other solution methods for problems with different numerical characteristics, then students may be better able to acknowledge that there is a need to study more than one solution method. In fact, it would be in the interest of the students to further recognize this utility to the extent that it will help them primarily to solve problems more competently and adaptively, and secondarily to gain exposure to and to practice satisfying their desire in their quest for the most "beautiful" solution methods.

Nonetheless, the order of presenting a series of different numerical characteristics of problems necessitates a careful deliberation of cognitive workload. In order for students to discover the power of certain solution methods, numerical characteristics of problems need to be reflected upon in a manner that unfolds the necessity of those solution methods. Correspondingly, the order of presenting the many different solution methods should be connected with the amount of cognitive workload demanded in each numerical characteristic of the problems in an increasing manner. As students grow their mathematical confidence in solving problems with numerical characteristics of lower cognitive workload, they may be introduced to problems with numerical characteristics of higher cognitive workload. With constant exposure to having to deal with problems with numerical characteristics of higher cognitive workload, students may come to realize that they need some other solution methods that are more effective than the existing solution methods that normally work just fine with problems with numerical characteristics of lower cognitive workload.

In many respects, this model to align itself closely with the kind of preferential considerations involved in the problem solving experience that professional mathematicians encounter in their research work when deciding which approaches to pursue in solving a theorem, which existing theorems to prove, or which new theorems to conjecture. Likewise, such considerations can be related in the review process in which mathematics textbook authors conduct in determining which proofs, among existing ones, to include for each theorem in the textbooks. Indeed, a survey of research in the field of mathematics reveals that there are more proofs than there are theorems (Thurston, 1994). This indicates that more and more mathematicians are working on to use different approaches and perspectives to revisit and refine many theorems that have already been proved.

At the heart of the present model for aesthetics in school mathematics is the dynamic cycle of the need for more effective solution methods through different numerical characteristics of problems with similar surface structures, the existence of a collection of different solution methods through students' constructivism or teachers' presentation, and the uniqueness in the selection of preferred solution methods. It is the element of the gradual progression in time and difficulty that helps run the engine of this model. Students who respond to the need to learn problem solving using many different approaches may grow to become aware of "beautiful" solution methods. Students who experience a huge and sudden jump in the level of cognitive workload may in turn feel unmotivated to appreciate the "beautiful" solution methods. By creating the necessary condition for students to explore multiple solution methods, we facilitate a learning environment where they can engage in classroom discourse to compare and contrast those solution methods, and eventually instill the feelings for aesthetics in their mathematics learning experience.

Aesthetics for a Lesson in One- and Two-digit Addition

This section offers a concrete model of how a lesson in one- and two-digit addition might look like. We begin our lesson in addition as a transition from a lesson in counting. As students become acquainted with counting from one, two, three, and so on, our first examples in one-digit addition involve adding by ones: $0 + 1 = 1$, $1 + 1 = 2$, $2 + 1 = 3$, $3 + 1 = 4$, $4 + 1 = 5$, $5 + 1 = 6$, $6 + 1 = 7$, $7 + 1 = 8$, $8 + 1 = 9$, and $9 + 1 = 10$ (see Figure 1). This set of problems is introduced first because it serves as a reminder to students that adding by ones is tantamount to listing or naming numerals sequentially. Different methods such as using manipulatives, counting via fingers, or talking aloud can also be utilized.

+	0	1	2	3	4	5	6	7	8	9
0	0	1	2	3	4	5	6	7	8	9
1	1	2	3	4	5	6	7	8	9	10
2	2	3	4	5	6	7	8	9	10	11
3	3	4	5	6	7	8	9	10	11	12
4	4	5	6	7	8	9	10	11	12	13
5	5	6	7	8	9	10	11	12	13	14
6	6	7	8	9	10	11	12	13	14	15
7	7	8	9	10	11	12	13	14	15	16
8	8	9	10	11	12	13	14	15	16	17
9	9	10	11	12	13	14	15	16	17	18

Figure 1. Adding whole numbers by ones.

Following adding by ones problems is an introduction to commutative property of addition. This includes problems such as $1 + 0 = 1$, $1 + 1 = 2$, $1 + 2 = 3$, $1 + 3 = 4$, $1 + 4 = 5$, $1 + 5 = 6$, $1 + 6 = 7$, $1 + 7 = 8$, $1 + 8 = 9$, and $1 + 9 = 10$ (see Figure 2). This set of problems establishes the foundation for future addition problems that students will encounter to the extent that it demonstrates the economy aspect of solution methods. Students will recognize that the order in which an addition is performed does not matter when adding two whole numbers: adding an addend to an augend is equivalent to adding an augend to an addend.

+	0	1	2	3	4	5	6	7	8	9
0	0	1	2	3	4	5	6	7	8	9
1	1	2	3	4	5	6	7	8	9	10
2	2	3	4	5	6	7	8	9	10	11
3	3	4	5	6	7	8	9	10	11	12
4	4	5	6	7	8	9	10	11	12	13
5	5	6	7	8	9	10	11	12	13	14
6	6	7	8	9	10	11	12	13	14	15
7	7	8	9	10	11	12	13	14	15	16
8	8	9	10	11	12	13	14	15	16	17
9	9	10	11	12	13	14	15	16	17	18

Figure 2. Adding ones to whole numbers.

Given the first two sets of problems, we can now reintroduce to students that whole numbers can be represented through decomposition as (at least) two other (not necessarily distinguishable) whole numbers. Students can be engaged to examine the question of which whole numbers can make up different whole numbers. For example, the number zero can be represented as $0 = 0 + 0$; the number one can be represented as $1 = 0 + 1$ and $1 = 1 + 0$; the number two can be represented as $2 = 0 + 2$, $2 = 1 + 1$, and $2 = 2 + 0$; the number three can be represented as $3 = 0 + 3$, $3 = 1 + 2$, $3 = 2 + 1$, and $3 = 3 + 0$; the number four can be represented as $4 = 0 + 4$, $4 = 1 + 3$, $4 = 2 + 2$, $4 = 3 + 1$, and $4 = 4 + 0$; the number five can be represented as $5 = 0 + 5$, $5 = 1 + 4$, $5 = 2 + 3$, $5 = 3 + 2$, $5 = 4 + 1$, and $5 = 5 + 0$; the number six can be represented as $6 = 0 + 6$, $6 = 1 + 5$, $6 = 2 + 4$, $6 = 3 + 3$, $6 = 4 + 2$, $6 = 5 + 1$, and $6 = 6 + 0$; the number seven can be represented as $7 = 0 + 7$, $7 = 1 + 6$, $7 = 2 + 5$, $7 = 3 + 4$, $7 = 4 + 3$, $7 = 5 +$

2, 7 = 6 + 1, and 7 = 7 + 0; the number eight can be represented as 8 = 0 + 8, 8 = 1 + 7, 8 = 2 + 6, 8 = 3 + 5, 8 = 4 + 4, 8 = 5 + 3, 8 = 6 + 2, 8 = 7 + 1, and 8 = 8 + 0; and the number nine can be represented as 9 = 0 + 9, 9 = 1 + 8, 9 = 2 + 7, 9 = 3 + 6, 9 = 4 + 5, 9 = 5 + 4, 9 = 6 + 3, 9 = 7 + 2, 9 = 8 + 1, and 9 = 9 + 0 (see Figure 3).

+	0	1	2	3	4	5	6	7	8	9
0	0	1	2	3	4	5	6	7	8	9
1	1	2	3	4	5	6	7	8	9	10
2	2	3	4	5	6	7	8	9	10	11
3	3	4	5	6	7	8	9	10	11	12
4	4	5	6	7	8	9	10	11	12	13
5	5	6	7	8	9	10	11	12	13	14
6	6	7	8	9	10	11	12	13	14	15
7	7	8	9	10	11	12	13	14	15	16
8	8	9	10	11	12	13	14	15	16	17
9	9	10	11	12	13	14	15	16	17	18

Figure 3. Adding through decomposition.

When faced with addition problems that result in one-digit whole numbers, students can apply their commutative property of addition. After some experience of using the counting on technique that starts either from the first addend or from the greater of the two addends, students can choose the latter as their preferred method because of its economical consideration.

The next set of problems involves whole numbers up to 18 that are the results of additional one-digit addends. For example, the number 10 can be represented as 10 = 1 + 9, 10 = 2 + 8, 10 = 3 + 7, 10 = 4 + 6, 10 = 5 + 5, 10 = 6 + 4, 10 = 7 + 3, 10 = 8 + 2, and 10 = 9 + 1; the number 11 can be represented as 11 = 2 + 9, 11 = 3 + 8, 11 = 4 + 7, 11 = 5 + 6, 11 = 6 + 5, 11 = 7 + 4, 11 = 8 + 3, and 11 = 9 + 2; the number 12 can be represented as 12 = 3 + 9, 12 = 4 + 8, 12 = 5 + 7, 12 = 6 + 6, 12 = 7 + 5, 12 = 8 + 4, and 12 = 9 + 3; the number 13 can be represented as 13 = 4 + 9, 13 = 5 + 8, 13 = 6 + 7, 13 = 7 + 6, 13 = 8 + 5, and 13 = 9 + 4; the number 14 can be represented as 14 = 5 + 9, 14 = 6 + 8, 14 = 7 + 7, 14 = 8 + 6, and 14 = 9 + 5; the number 15 can be represented as 15 = 6 + 9, 15 = 7 + 8, 15 = 8 + 7, and 15 = 9 + 6; the number 16 can be represented as 16 = 7 + 9, 16 = 8 + 8, and 16 = 9 + 7; the number 17 can be represented as 17 = 8 + 9, and 17 = 9 + 8; and the number 18 can be represented as 18 = 9 + 9 (see Figure 4).

+	0	1	2	3	4	5	6	7	8	9
0	0	1	2	3	4	5	6	7	8	9
1	1	2	3	4	5	6	7	8	9	10
2	2	3	4	5	6	7	8	9	10	11
3	3	4	5	6	7	8	9	10	11	12
4	4	5	6	7	8	9	10	11	12	13
5	5	6	7	8	9	10	11	12	13	14
6	6	7	8	9	10	11	12	13	14	15
7	7	8	9	10	11	12	13	14	15	16
8	8	9	10	11	12	13	14	15	16	17
9	9	10	11	12	13	14	15	16	17	18

Figure 4. Additional additions through decomposition.

Using the knowledge of commutative property of addition, students need to be prompted to realize that from the last two sets of the problems, many decompositions are symmetric to each other. Moreover, some students who may recognize that the number of decompositions of a whole number is one more than that whole number can be guided to think about whether this observation holds for the cases for 10 to 18 shown previously. Students need to see that one-digit addition problems will result between zero and 18. More specifically, they should also notice that the maximum sum of two one-digit whole numbers is 18, a useful fact that will later be employed in the traditional vertical algorithm as the standard algorithm.

At this point, we expand our addition problems using two-digit whole numbers as addends. For example, in addition to the decompositions shown previously, the number 10 can be represented as 10 = 0 + 10, and 10 = 10 + 0; the number 11 can be represented as 11 = 0 + 11,

11 = 1 + 10, 11 = 10 + 1, and 11 = 11 + 0; the number 12 can be represented as 12 = 0 + 12, 12 = 1 + 11, 12 = 2 + 10, 12 = 10 + 2, 12 = 11 + 1, and 12 = 12 + 0; the number 13 can be represented as 13 = 0 + 13, 13 = 1 + 12, 13 = 2 + 11, 13 = 3 + 10, 13 = 10 + 3, 13 = 11 + 2, 13 = 12 + 1, and 13 = 13 + 0; the number 14 can be represented as 14 = 0 + 14, 14 = 1 + 13, 14 = 2 + 12, 14 = 3 + 11, 14 = 4 + 10, 14 = 10 + 4, 14 = 11 + 3, 14 = 12 + 2, 14 = 13 + 1, and 14 = 14 + 0; the number 15 can be represented as 15 = 0 + 15, 15 = 1 + 14, 15 = 2 + 13, 15 = 3 + 12, 15 = 4 + 11, 15 = 5 + 10, 15 = 10 + 5, 15 = 11 + 4, 15 = 12 + 3, 15 = 13 + 2, 15 = 14 + 1, and 15 = 15 + 0; the number 16 can be represented as 16 = 0 + 16, 16 = 1 + 15, 16 = 2 + 14, 16 = 3 + 13, 16 = 4 + 12, 16 = 5 + 11, 16 = 6 + 10, 16 = 10 + 6, 16 = 11 + 5, 16 = 12 + 4, 16 = 13 + 3, 16 = 14 + 2, 16 = 15 + 1, and 16 = 16 + 0; the number 17 can be represented as 17 = 0 + 17, 17 = 1 + 16, 17 = 2 + 15, 17 = 3 + 14, 17 = 4 + 13, 17 = 5 + 12, 17 = 6 + 11, 17 = 7 + 10, 17 = 10 + 7, 17 = 11 + 6, 17 = 12 + 5, 17 = 13 + 4, 17 = 14 + 3, 17 = 15 + 2, 17 = 16 + 1, and 17 = 17 + 0; and the number 18 can be represented as 18 = 0 + 18, 18 = 1 + 17, 18 = 2 + 16, 18 = 3 + 15, 18 = 4 + 14, 18 = 5 + 13, 18 = 6 + 12, 18 = 7 + 11, 18 = 8 + 10, 18 = 10 + 8, 18 = 11 + 7, 18 = 12 + 6, 18 = 13 + 5, 18 = 14 + 4, 18 = 15 + 3, 18 = 16 + 2, 18 = 17 + 1, and 18 = 18 + 0 (see Figure 5).

+	0	1	2	3	4	5	6	7	8	9	10	11	12	13	14	15	16	17	18
0	0	1	2	3	4	5	6	7	8	9	10	11	12	13	14	15	16	17	18
1	1	2	3	4	5	6	7	8	9	10	11	12	13	14	15	16	17	18	19
2	2	3	4	5	6	7	8	9	10	11	12	13	14	15	16	17	18	19	20
3	3	4	5	6	7	8	9	10	11	12	13	14	15	16	17	18	19	20	21
4	4	5	6	7	8	9	10	11	12	13	14	15	16	17	18	19	20	21	22
5	5	6	7	8	9	10	11	12	13	14	15	16	17	18	19	20	21	22	23
6	6	7	8	9	10	11	12	13	14	15	16	17	18	19	20	21	22	23	24
7	7	8	9	10	11	12	13	14	15	16	17	18	19	20	21	22	23	24	25
8	8	9	10	11	12	13	14	15	16	17	18	19	20	21	22	23	24	25	26
9	9	10	11	12	13	14	15	16	17	18	19	20	21	22	23	24	25	26	27
10	10	11	12	13	14	15	16	17	18	19	20	21	22	23	24	25	26	27	28
11	11	12	13	14	15	16	17	18	19	20	21	22	23	24	25	26	27	28	29
12	12	13	14	15	16	17	18	19	20	21	22	23	24	25	26	27	28	29	30
13	13	14	15	16	17	18	19	20	21	22	23	24	25	26	27	28	29	30	31
14	14	15	16	17	18	19	20	21	22	23	24	25	26	27	28	29	30	31	32
15	15	16	17	18	19	20	21	22	23	24	25	26	27	28	29	30	31	32	33
16	16	17	18	19	20	21	22	23	24	25	26	27	28	29	30	31	32	33	34
17	17	18	19	20	21	22	23	24	25	26	27	28	29	30	31	32	33	34	35
18	18	19	20	21	22	23	24	25	26	27	28	29	30	31	32	33	34	35	36

Figure 5. Additional additions through decomposition of two-digit whole numbers.

After a systematic introduction to addition problems involving whole numbers between zero and 18, students can be further exposed to more complex two-digit whole numbers. For problems involving two-digit whole numbers with zero in their ones place values, we begin with the strategy of counting on by tens. For example, 10 + 10 = 20, 20 + 10 = 30, 30 + 10 = 40, 40 + 10 = 50, 50 + 10 = 60, 60 + 10 = 70, 70 + 10 = 80, 80 + 10 = 90, and 90 + 10 = 100. This set of problems should remind students of that involving counting on by ones technique. Students can further extrapolate adding two-digit whole numbers with zero in their ones place values has a similar pattern that one can find when adding one-digit whole numbers.

Moreover, combining the two techniques will enable students to further generalize the pattern in adding any two-digit whole numbers: when adding two-digit whole numbers, one can add the digits in the tens place values, add the digits in the ones place values, and combine the two summands. For example, $34 + 52 = (30 + 4) + (50 + 2) = (30 + 50) + (4 + 2) = 80 + 6 = 86$, and $46 + 38 = (40 + 6) + (30 +8) = (40 + 30) + (6 + 8) = 70 + 14 = 70 + (10 + 4) = (70 + 10) + 4 = 80 + 4 = 84$. Students need to notice that the latter example involves a higher level of cognitive workload than the former one because of the resulting two-digit summand from adding the digits in the ones place values. As such, we need to be more cognizant of presenting addition problems of two-digit whole numbers so that they will progress from those involving a one-digit summand to those involving a two-digit summand as a result of adding the digits in the ones place values.

We make a note at this point that although the previous technique of adding any two-digit whole numbers involves associative property of addition, the introduction to this property is rather informal to our first grade students. In fact, students may choose not to write formal number sentences to express any of the solution methods that are described above. Formality or standardization will nevertheless become necessary for students to adapt to as they are posed with more complex addition problems such as those involving more than two two-digit whole numbers (or even those involving two more-than-two-digit whole numbers as they will perform in later grade levels).

Consequently, students can be introduced to the traditional vertical algorithm, which essentially reverse the order of the previous technique in a vertical form. For example, $46 + 38 = (40 + 6) + (30 + 8) = (40 + 30) + (6 + 8) = (40 + 30) + (14) = (40 + 30) + (10 + 4) = (40 + 30 +$

10) + 4 = 80 + 4 = 84. We point out to students that the step that involves adding 10 to 40 + 30 is referred to as carrying over, and that 10 is the only possible value to carry over because of the observation made earlier that the maximum sum of two one-digit whole numbers is 18. With this traditional vertical algorithm in their tool bag, students are hoped to recognize its advantage over other solution methods in terms of generalizability: this algorithm, unlike the others, will work for addition problems involving not only two one- or two-digit whole numbers, but also more than two one- or two-digit whole numbers and two more-than-two-digit whole numbers.

Some other students might call attention to its advantage in terms of efficiency in the time that it takes to solve the addition problems. To these students, we propose a situation for addition problems: Is there a more efficient way of solving the addition problem such as 46 + 99 than using the traditional vertical algorithm? Students might realize that decomposing 46 into 45 and that 1 will make up 100 from 1 and 99, which is quicker to solve than the decomposition method: 46 + 99 = (45 + 1) + 99 = 45 + (1 + 99) = 45 + 100 = 145. Alternatively, students might recognize that compensating 99 into 100 and that 1 will need to be taken away from 46 and 100, which is again quicker to solve than the decomposition method: 46 + 99 = 46 + (100 – 1) = (46 + 100) – 1 = 146 – 1 = 145 Clearly, there exists a situation where the traditional vertical algorithm may not always offer a faster solution than the decomposition method or the compensation method. And for students to understand the need for flexibility in applying more effective solution methods to appropriate addition problems, extraordinary situations as described earlier may be indispensible.

Perhaps even generalizing the decomposition and compensation methods a little further to addition problems involving more than two two-digit whole numbers, students can recognize that some digits (whole numbers) are more compatible to other digits (whole numbers) by taking advantage of the simplicity of multiple of tens. For example, as students recognize that the digits 6 and 4 make up 10, they can associate the corresponding addends using the commutative and associative properties of addition: $46 + 38 + 54 = 46 + (38 + 54) = 46 + (54 + 38) = (46 + 54) + 38 = 100 + 38 = 138$.

Although first graders are only expected to perform addition to problems involving two whole numbers up to 100 (Common Core State Standards Initiative, 2010), it is perhaps worth contemplating to what extent the power and beauty of the traditional vertical algorithm as the standard algorithm, as well as some other solution methods such as decomposition or compensation method, may become unnoticed, if not optimally appreciated, in the absence of problems requiring a higher level cognitive workload. It is through this continuous process of experiencing such problems with a variety of numerical characteristics that students can become aware of the need for more effective solution methods. And it is through this need for additional solution methods that student can not only engage in the creative praxis of constructing and inventing their own solution methods, but also learn to acknowledge the simultaneous existence of multiple solution methods. It is then through this existence of many different solution methods that students can grow to analyze those solution methods more critically toward the aesthetic goals of learning in mathematics.

Conclusion and Discussion

This article presents a case of aesthetics for school mathematics. In an expository approach, it aims to illustrate the possibility that aesthetics may find its presence in school mathematics through problem solving. We recognize that mathematical aesthetics may not be necessarily interpreted by the sole means of the process in which mathematicians go through in their professional career (Poincare, 1946; Hadamard, 1945). Our approach to interpret mathematical aesthetics through problem solving is mostly influenced by the eminent call to teach mathematics through problem solving (NCTM, 2000). Using the model proposed, we consider the necessity, existence, and uniqueness of mathematically "beautiful" solution methods in student learning. It is hoped that this proposed model might create discussions in advancing research in teaching and learning mathematics.

On the one hand, the proposed model approximates the role of generating mathematical understanding from the point of view of many of the existing teaching and learning theories in mathematics education (Cobb, 2007; Hiebert & Carpenter, 1992; Lampert, 1990; Silver & Herbst, 2007; Simon, 1995). On the other hand, the proposed model operates mathematical aesthetics around the classroom settings where mathematics is conducted in the lens of abstract problem solving process as observed in the studies by Hadamard (1945) and Poincare (1946), instead of through the concrete and static appearance of problem solving solutions as observed in the studies by Krutetskii (1976) and Sinclair (2001). At the same time, our approach might be likened to constructivism to the extent that our aesthetic engagement process promotes creativity and analytical thinking through a series of solution methods and their corresponding numerical characteristics of problems with a similar surface structure. In contrast to the current curriculum, not only does this model support students' accumulation of a generally more substantial problem solving experience, but it also incorporates acquisition of new knowledge through the creation, presence, and evaluation of many different strategies.

The present article also demonstrates one possibility of engaging first grade students aesthetically in mathematics learning of the one- and two-digit addition. A lesson specifically considers a greater amount of depth and breath of the treatment of numerical characteristics of the addends involved in the addition problems in manner that incorporates systematically and progressively increasing cognitive workload. Accordingly, this model has a number of implications that are worth of some reflections. First, the exposition of a substantially greater series of addition problems, although valuable in creating the need for searching different problem solving approaches, should be ensured not to exhaust the mathematical excitement of the students. Teachers should be cognizant of how much exertion in the planning of a series of problems in the addition lesson may be sufficient over a given period of time. This circumstance may also create a glimpse of constriction among other topics that can be taken account of under a particular grade level. It is conceivable that certain methods of algebra are capable of solving typical calculus problems (Tjoe, 2015). Teachers need to strike a balance in maintaining how far students can or should invoke some of the most powerful approaches in algebra class to solve problems involving a lower level of cognitive workload such as those that one can find in an elementary class, or involving a higher level of cognitive workload such as those that one can find in an AP Calculus or an advanced geometry class. Finally, future studies are called for to attend to empirical findings that support the proposed model of aesthetics engagement in school mathematics (Tjoe, 2014). Such studies may particularly relate to an experiment that compares and contrasts the efficacy of such variables as numerical characteristics of problems and order of presentation of different solution methods.

References

Cobb, P. (2007). Putting philosophy to work: Coping with multiple theoretical perspectives. In F. K. Lester (Ed.), *Second handbook of research on mathematics teaching and learning* (pp. 3-38). Charlotte, NC: Information Age.

Common Core State Standards Initiative. (2010). *Common core state standards for mathematics.* Retrieved from http://www.corestandards.org/assets/CCSSI_Math%20Standards.pdf

Davis, P.J. and Hersh, R. (1981). *The mathematical experience*. Boston, MA: Birkhauser.

Dreyfus, T., & Eisenberg, T. (1986). On the aesthetic of mathematical thought. *For the Learning of Mathematics*, *6*, 2-10.

Geary, D. C., & Brown, S. C. (1991). Cognitive addition: Strategy choice and speed-of-processing differences in gifted, normal, and mathematically disabled children. *Developmental Psychology*, *27*, 398-406.

Geary, D. C., & Wiley, J. G. (1991). Cognitive addition: Strategy choice and speed-of-processing differences in young and elderly adults. *Psychology and Aging*, *6*, 474-483.

Hadamard, J. (1945). *The psychology of invention in the mathematical field.* Princeton, NJ: Princeton University Press.

Hardy, G. H. (1940). *A mathematician's apology*. Cambridge, England: Cambridge University Press.

Hiebert, J., & Carpenter, T. P. (1992). Learning and teaching with understanding. In D. A. Grouws (Ed.), *Handbook of research on mathematics teaching and learning* (pp. 65–97). New York, NY: Macmillan.

Karp, A. P. (2008). Which problems do teachers consider beautiful? A comparative study. *For the Learning of Mathematics*, *28*, 36-43.

Krutetskii, V. A. (1976). *The psychology of mathematical abilities in schoolchildren* (J. Kilpatrick, I. Wirszup, Eds., & J. Teller, Trans.). Chicago, IL: University of Chicago Press.

Lampert, M. (1990). When the problem is not the question and the solution is not the answer: Mathematical knowing and teaching. *American Educational Research Journal*, *27*, 29-63.

Papert, S. (1980). *Mindstorms: Children, computers, and powerful ideas*. New York, NY: Basic Books.

Poincare, H. (1946). *The foundations of science* (G. B. Halsted, Trans.). Lancaster, PA: Science Press.

Siegler, R. S., & Shrager, J. (1984). Strategy choices in addition and subtraction: How do children know what to do? In C. Sophian (Ed.), *The origins of cognitive skills* (pp. 229-293). Hillsdale, NJ: Erlbaum.

Silver, E. A., & Herbst, P. G. (2007). Theory in mathematics education scholarship. In F. K. Lester (Ed.), *Second handbook of research on mathematics teaching and learning* (pp. 39-68). Charlotte, NC: Information Age.

Silver, E. A., & Metzger, W. (1989). Aesthetic influences on expert mathematical problem solving. In D. McLeod & V. Adams (Eds.), *Affect and mathematical problem solving* (pp. 59-74). New York, NY: Springer-Verlag.

Simon, M. A. (1995). Reconstructing mathematics pedagogy from a constructivist perspective. *Journal for Research in Mathematics Education*, *26*, 114-145.

Sinclair, N. (2001). The aesthetic *is* relevant. *For the Learning of Mathematics*, *21*, 25-32.

Sinclair, N. (2004). The roles of the aesthetic in mathematical inquiry. *Mathematical Thinking and Learning, 6*, 261-284.

Smith, D. E. (1927). Esthetics and mathematics. *The Mathematics Teacher, 20*, 419-428.

Sriraman, B. (2009). The characteristics of mathematical creativity. *ZDM- The International Journal on Mathematics Education, 41*, 13-27.

Thurston, W. P. (1994). On proof and progress in mathematics. *Bulletin of the American Mathematical Society, 20*, 161-177.

Tjoe, H. (2014). When understanding evokes appreciation: The effect of mathematics content knowledge on aesthetic predisposition. In C. Nicol, S. Oesterle, P. Liljedahl, & D. Allan. (Eds.), *Proceedings of the 38th Conference of the International Group for the Psychology of Mathematics Education* (Vol. 5, pp. 249-256). Vancouver, BC: PME.

Tjoe, H. (2015). Giftedness and aesthetics: Perspectives of expert mathematicians and mathematically gifted students. *Gifted Child Quarterly, 59*, 165-176.

Plato on the foundations of Modern Theorem Provers

Inês Hipolito[1]
Nova University of Lisbon

Abstract: Is it possible to achieve such a proof that is independent of both acts and dispositions of the human mind? Plato is one of the great contributors to the foundations of mathematics. He discussed, 2400 years ago, the importance of clear and precise definitions as fundamental entities in mathematics, independent of the human mind. In the seventh book of his masterpiece, *The Republic*, Plato states "arithmetic has a very great and elevating effect, compelling the soul to reason about abstract number, and rebelling against the introduction of visible or tangible objects into the argument" (525c). In the light of this thought, I will discuss the status of mathematical entities in the twentieth first century, an era when it is already possible to demonstrate theorems and construct formal axiomatic derivations of remarkable complexity with artificial intelligent agents — the *modern theorem provers*.

Keywords: Plato; Modern Theorem Provers; Formal Proof.

[1] hipolito.ines@gmail.com

Introduction

According to Platonism, a mathematical proof is the metaphysical view that there are abstract mathematical objects whose existence is independent of human language, thought, and practices. A *full-blooded Platonism* or *platitudinous Platonism* (FBP) asserts that it is possible for human beings to have systematically and non-accidentally true beliefs about a platonic mathematic realm — a mathematical realm satisfying *Existence*, *Abstractness* and *Independence*. Could there be such proof? Could a proof be objective and completely understood, independently of the possibilities of our knowing of truth or falsity?

Plato is one of the great contributors to the foundations of mathematics. He discussed, 2400 years ago, the importance of clear and precise definitions as fundamental entities in mathematics. In the seventh book of his masterpiece, *The Republic*, Plato states "arithmetic has a very great and elevating effect, compelling the soul to reason about abstract number, and rebelling against the introduction of visible or tangible objects into the argument". In the light of this thought, I will discuss the status of mathematical entities in the twentieth first century, an era where it is already possible to demonstrate theorems, construct formal axiomatic derivations of remarkable complexity with artificial intelligent agents — *the modern theorem provers*. A computer-assisted proof is written in a precise artificial language that admits only a fixed repertoire of stylized steps. It is formalized through artificial intelligent agents that mechanically verify, in a formal language, the correctness of the proof previously demonstrated by the human mind.

In contrast, calculi are exactly the kind of feature, which make it appealing for mathematicians. There are two reasons for this: (i.) it can be studied for its own properties and elegance of pure mathematics; and (ii.) can easily be extended to include other fundamental aspects of reasoning. According to Hofstadter, (1979), a proof is something informal, that is, a product of human thought, written in human language for human consumption. All sorts of complex features of thought may be used in proofs, and, thought they may "feel right", one may wonder if they can be logically defended. This is really what formalization is for.

1. Plato's conception of Arithmetic

According to Plato, at the end of the sixth book of *The Republic*, mathematicians' method of thinking is not a matter of intelligence, but rather a matter of *διανοια,* which means *understanding*. This is a definition by Plato that seems to etimologically imply δια (*between*), *νουσ* (*intelligence*) and *δοξα* (*opinion*), as if understanding would be something in between *opinion* and *inteligence*.

In 525a, Plato considers the concept of number, as a non-limited unity trough plurality, since "this characteristic occurs in the case of one; for we see the same thing to be both one and infinite in multitude" (525a). Also, with this conception of plurality as much as unity "thought begins to be aroused within us, and the soul perplexed and wanting to arrive at a decision asks 'What is absolute unity?' This is the way in which the study of the one has a power of drawing and converting the mind to the contemplation of reality." (525a).

As reported by Plato, the reality of calculus is a pure contemplation since in reasoning about numbers there are no visible bodies:

> "Plato — Now, suppose a person were to say to them, Glaucon, 'O my friends, what are these wonderful numbers about which you are reasoning, in which, as you say, there are constituent units, such as you demand, and each unit is equal to every other, invariable, and not divisible into parts,' - what would they answer?
>
> Glaucon —They would answer, as I should think, that they were speaking of those numbers which can only be realized in thought, and there is no other way of handling them." (*Republic*, 526a).

This means that arithmetic compels the soul to reach the pure truth trough intelligence. Furthermore, Plato considered the idea of *good* to be the ultimate objective of philosophy: "in the world of knowledge the idea of *good* appears last of all, and is seen only with an effort; and, when seen, is also inferred to be the universal author of all things beautiful and right" (526d). In his perspective, to accomplish the ideal of *good*, it is necessary that one study arithmetic and geometry, since they have two important characteristics. First, they invite thought and lead the mind to reflect and, accordingly, they allow the mind to grasp truth. Second, the advanced parts of mathematics and geometry have the power to draw the soul from becoming to beings: the true use of arithmetic. Therefore, the easiest way for the soul to go from becoming is to pursue the study of arithmetic until one is able to see the natures of numbers with the mind only. Moreover, "arithmetic has a very great and elevating effect, compelling the soul to reason about abstract number and repelling against the introduction of visible or tangible objects into the argument" (525c).

In agreement with Plato's considerations, mathematics has a philosophical importance since mathematics is a tool that helps and exercises the mind to think. This training process will lead to a better understanding and to the accomplishment of the idea of *good*, which is the crucial purpose in Philosophy. Indeed, the main philosophical importance of mathematics is the rewarding it may have in one's understanding of the reality. This is possible because mathematical objects are Forms: they can be completely separated from perceptible objects and they exhibit the same traits that philosophical objects. This means that mathematical objects are not grasped by the senses but by the intellect. In *The Republic*, on the one hand, mathematical objects are intelligible and can be known and, on the other, mathematical axioms are accepted as true without further proof or investigation. This view stands that unlike physical objects, mathematical objects do not exist in space and time; rather there are abstract mathematical objects whose existence is independent of the human mind: just as electrons and planets exist independently of us, so do numbers and sets. And just as statements about electrons and planets are made true or false by the objects with which they are concerned, so are statements about numbers and sets. Correspondingly, one may say that mathematical Platonism may be defined as a conjunction of three theses: *Existence*, *Abstractness,* and *Independence*.

3. Existence, Abstractness and Independence

According to Dummett, 1978, Platonism, as a philosophy of mathematics, is founded on a simile: the comparison between the apprehension of mathematical truth to the perceptions of physical objects, and thus of mathematical reality to the physical universe. Gödel (1995), asserts that Platonism is the view that mathematics describes a non-sensual reality, which exists independently both of acts and [of] the dispositions of the human mind and is the only perceived, and probably perceived very incompletely, by the human mind.

Maddy (1990), stresses that realism of Platonism is the view that mathematics is the scientific study of objectively existing mathematical entities just as physics is the study of physical entities. The statements of mathematics are true or false depending on the properties of those entities, independent of our ability, or lack thereof, to determine which. Parsons (1983), sustains that Platonism means here not just accepting abstract entities or universals but epistemological or metaphysical realism with respect to them. Thus, a platonistic interpretation of a theory of mathematical objects takes the truth or falsity of statements of the theory, in particular statements of existence, to be objectively determined independently of the possibilities of our knowing this truth or falsity.

There are several theses within the mathematical Platonism such as (1) Frege's arithmetic-object argument; (2) Quine's; and (3) a response that is commonly referred as full-blooded or platitudinous Platonism[2] (FBP).

The fundamental idea behind FBP is that it is possible for human beings to have systematically and non-accidentally true beliefs about platonic mathematical realm — a mathematical realm satisfying the *Existence*, *Abstractness* and *Independence* — without that realm in any way influencing us or us influencing it. These three thesis are made possible in virtue of (i.) Schematic Reference: the reference relation between mathematical theories and the mathematical realm is purely schematic, or at least close to purely schematic and (ii.) Plenitude: the mathematical realm is very large, it contains entities that are related to one another in all of the possible ways that entities can be related to one another.

Moreover, *Existence*, *Abstractness* and *Independence* seem to be validated by two other thesis (i.) mathematical theories embed collections of constraints on what the ontological structure of a given "part" of the mathematical (ii.) The existence of *any* such appropriate part of the mathematical realm is sufficient to make the said theory true of that part of that realm. In agreement with Roger Penrose,

> Platonic existence, as I see it, refers to the existence of an objective external standard that is not dependent upon our individual opinions nor upon our particular culture. Such 'existence' could also refer to things other than mathematics, such as to morality or aesthetics, but I am here concerned just with mathematical objectivity, which seems to be a much clearer issue...
>
> Plato himself would have insisted that there are two other fundamental absolute ideals, namely that of the *Beautiful* and that of the *Good*. I am not at all adverse to admitting the existence of such ideals, and to allowing the Platonic world to be extended so as to contain absolutes of this nature (Penrose, 2007).

---

[2] This third response has been most fully articulated by Mark Balaguer (1998) and Stewart Shapiro (1997).

The concept *Existence* in mathematical objects is rather controversial. If this view is true, it will dig up the physicalist idea that the reality is exhausted by the physical and will also put great pressure on many naturalistic theories of knowledge[3], since there is little doubt that the human mind possesses mathematical knowledge. Burgess has defended anti-nominalism. *Anti-nominalism* is, simply, the rejection of nominalism. As such, anti-nominalists endorse ontological commitment to mathematical entities, but refuse to engage in speculation about the metaphysical nature of mathematical entities that goes beyond what can be supported by common sense and science (Burgess 1983, and Burgess and Rosen 1997, 2005). Anti-nominalism, is the conjunction of *Existence* and *Abstractedness*, and its consequences are not as strong as of Platonism. Some views such as intuitionistic, are anti-nominalistic about being a Platonism, for the existence of mathematical objects are true however, these objects depend on or are constituted by mathematicians.

On the *Truth-value realism* perspective[4], Mathematical objects must exist since they are true of all. Mathematical objects have the existence in virtue of its unique and objective truth-value independent of whether we can know it and whether it follows logically from our current mathematical theories. Mathematical objects have a unique and objective truth-value. Expressly, "a sentence proper is a proper name, and its *Bedeutung*[5], if it has one, is a truth-value: the True or the False" (Beaney 1997, 297). This is clearly a metaphysical view, but not an ontological view, as Platonism, since truth-values are not committed to the flow from an ontology that demands *Existence*.

For another hand, there is the *working realism* methodological view that entails mathematics should be practiced as if Platonism was true (Bernays, 1935, Shapiro, 1997), considering that working realism is first and foremost a view within mathematics itself about the correct methodology of the discipline, and Platonism, rather, is a philosophical view. Nevertheless, as reported by Hoystein (2013), this two are very related theories since, assuming Platonism is true, then (1) language of mathematics is a classical first-order language; (2) provided it is legitimate to reason classically about any independently existing part of reality, classical rather than intuitionistic logic; (3) since Platonism ensures that mathematics is discovered rather than invented, there would be no need for mathematicians to restrict themselves to constructive methods and axioms, which establishes non-constructive methods; (4) impredicative definitions are legitimate whenever the objects being defined exist independently of our definitions, and this is how we assure the impredicative definitions; (5) if mathematics is about some independently existing reality, then every mathematical problem has a unique and determinate answer, which provides at least some motivation for Hilbertian optimism. As we can see, the truth of mathematical Platonism has important consequences within mathematics itself.

On this matter, Frege developed an argument — the *Fregean argument* — which is based on two premises: (1) The singular terms of the language of mathematics purport to refer to mathematical objects, and its first-order quantifiers purport to range over such

[3] In the nominalist perspective there exist no abstract or universal objects: this position denies the existence of abstract objects – objects that do not exist in space and time.

[4] The notion of a *truth value* has been explicitly introduced into logic and philosophy by Gottlob Frege—for the first time in 1891, and most notably in his seminal paper (1892).

[5] *meaning*

objects, this is, *S* to be true, must succeed in referring or quantifying, a classical semantics perspective; (2) most sentences accepted as mathematical theorems are true (regardless of their syntactic and semantic structure). By *Classical Semantics* and *Truth*, Frege considers that some simple numerical identities are objectively true because such identities allow the application of natural numbers in representing and reasoning about reality, especially the non-mathematical parts of reality. Other versions of Fregean argument are sometimes stated as the notion of ontological commitment, such as the Quine's Criterion[6].

On the other hand, *Abstractness* asserts that a mathematical object is said to be abstract just in case it is non-spatiotemporal, and therefore causally inefficacious. In Plato's thought, this distinction embodies the distinction between Forms, and Sensibles. In the seventeenth century, Locke's idea of an abstract object — as one that is formed from concrete ideas — was rejected by Berkeley and then by Hume. However, even for Locke there was no suggestion that the distinction between abstract ideas and concrete corresponded to a distinction among objects. In the twentieth century, Frege insisted that the objectivity and aprioricity of the truths in mathematics entail that numbers are neither material beings nor ideas in the mind. In The *Foundations of Arithmetic* (1974), Frege concludes that numbers are neither external "concrete" things nor mental entities of any sort. Later in *The Thought* (1952) he asserts that thoughts belong to a "third realm". Similar claims have been made by Bolzano, and later by Brentano, and his pupil Husserl.

In most recent attempts to distinguish concrete objects from abstract, Putnam (1975) makes the case for abstract objects on scientific grounds. Bealer (1993) and Tennant (1997) present *a priori* argument for the necessary existence of abstract entities. The dispute over the existence of *abstracta* is reviewed in Burges and Rosen (1997). A general theory of abstract objects is developed in Zalta (1983, 1999).

Independence is less evident than the other two claims. What does it mean for an object to be independent? Does it mean it is self-representative? An object seems to be independent when what it represents does not depend on any intelligent agent, or an agent's thought, reason, cognition or representation.

So far one may say that we have understood how mathematical objects assure their *Existence* and their *Abstractedness*. However, are mathematical entities sufficiently true to claim independence? Is it possible to entail mathematical objects as an objective concept? If *Independence* means that (1) mathematical objects are mind-independent; that (2) reality includes objects not subject to intentionality; that (3) to be referred requires a definition of truth; and that (4) there is only one correct description of the reality, then the trivial forms of Platonism are likely to satisfy the claim, and thus qualify Platonism with the property of *Independence.*

On this account, there are at least some objects, in reality, that could be perceived by the human mind via definition. On this assumption, an object is said to be independent when it is self-representative. Furthermore, we may also think of some other examples of self-representation. The Epimenides' paradox "All Cretans are liers" therefore, "I am a liar", *ergo* "This statement is false". This paradox relies on some form of self-reference. Any language

[6] A first-order sentence (or collection of such sentences) is ontologically committed to such objects as must be assumed to be in the range of the variables for the sentence (or collection of sentences) to be true.

capable of expressing some basic syntax can generate self-referential sentences. A language containing a truth predicate and this basic syntax will thus have a sentence L such that L implies $\neg Tr(\ulcorner L \urcorner)$ and vice versa.

Also Gödel's incompleteness theorems use mathematical reasoning in exploring mathematical reasoning itself. What the theorem states and how it is proved are, as a matter of fact, two different things. The theorem asserts that all consistent axiomatic formulation of number theory includes *undecidable* propositions. In the same way that Epimenides' paradox is a self-referential, also Gödel's axiom is a self-referential mathematics statement[7]. In Gödel–numbering, numbers are made to stand for symbols and sequences of symbols. Transporting the Epimenides' Paradox into number-theoretical formalism, the Epimenides' Paradox does not say

This statement of number-theory is false.

but that

This statement of number-theory does not have any proof.

We could agree that proofs are demonstrations within fixed systems of propositions. However, this statement of number theory does not have any proof in the system. Gödel's sentence is *unprovable* within the system, for there are true statements of number theory, which its methods of proof are too weak to demonstrate in the system; therefore, the system is incomplete. In fact, what Gödel showed was that provability is a weaker notion than truth no matter what axiomatic system is involved, and therefore, no fixed system could represent the complexity of the whole numbers[8]. Actually, if a system, such as the one defined in *Principia Mathematica*, is (i.) consistent, this is, *contradiction free*; and (ii.) complete, this is, every true statement of number theory could be derived within the framework drawn up in the P.M., how would it be possible to justify the methods of reason on the bases of that same methods of reasoning? If such a proof could be found using only methods inside a system then, the system itself would be inconsistent.

As we discussed on the first part of this essay, Plato considered the knowledge of arithmetic a *conditio* for philosophical knowledge since both require universal truths, accessible to the human mind by acquaintance to the incorporeal intelligible realm, — in opposition of *doxai*, which belong to a sensible sphere, — in resolving the problem of reality, knowledge and human existence.

Plato, in his considerations on arithmetic and geometry assesses epistemological issues, such as what and how the mind know things; and metaphysics, on their ontological status as things. As we discussed, a mathematical proof is a mathematical object that qualifies to the properties of *Existence*, *Abstractedness* and *Independence*. A proof is, a consequence of human reasoning, written in human language for human consumption. What is still left to

[7] For more detail on this issue, see Hofstadter (1979).
[8] Cf. Nagel and Newman, (2001).

discuss is whether a computer assisted proof — a proof that results from an artificial intelligent agent — may or may not qualify to those properties.

3. Modern Theorem Provers

A computer assisted proof is a proof in which every logical inference has been checked all the way back to the fundamentals axioms of mathematics (Hales, 2008). It is written in a precise artificial language that admits only a fixed repertoire of stylized steps (Harrison, 2008).

Proof assistants or *computer theorem provers* are the artificial intelligent agents that mechanically verify, in a formal language, the correctness of the proof previously demonstrated by the human mind. In fact, with this artificial system, the user is allowed to set up a mathematical theory, define properties and realize logical reasoning (Geuvers, 2009).

Nowadays, there is a large number of computer provers, that can check or construct computer assisted proofs, such as the HOL Light for classical and higher order logic, based on a formulation of *type theory*; and Coq. The theorem provers allow the expression of mathematical assertions; mechanically checks proofs; helps to find computer assisted proofs; and extracts a certified program from the constructive proof of its formal specification. Other theorem provers are Mizar, PVS, Otter/Ivy, Isabelle/Isar, Affa/Agda, ACL2, PhoX, IMPS, Metamath, Theorema, Lego, Nuprl, Ωmega, B method and Minlog.

According to Wiedijk (2006), theorem provers:

- are designed for the formalization of mathematics, or, if not designed specifically for that, have been seriously used for this purpose in the past;
- are special at something. These are the systems that in at least one dimension are better than all the other systems in the collection. They are the leaders in the field.

In recent years, several theorems have been formally verified by an artificial intelligent agent. There are many to choose from, however, some significant ones are:

(1) The *Four Colour Theorem*, states that, given any separation of a plane into contiguous regions, producing a figure called a *map*, no more than four colors are required to color the regions of the map so that no two adjacent regions have the same color. The theorem was proven in the late 19th century (Heawood 1890); however, proving that four colors suffice turned out to be pointedly harder. A number of false proofs and false counterexamples have appeared since the first statement of the four color theorem in 1852. Nevertheless, it was proven, using a computer, in 1976 by Kenneth Appel and Wolfgang Haken. It was the first major theorem to be proved using a computer. In 2005, Benjamin Werner and Georges Gonthier formalized a proof of the theorem inside the Coq proof assistant.

(2) The *Jordan Curve Theorem*, states that every simple closed curve in the plane separates the complement into two connected nonempty sets: an interior region and an exterior. The Jordan curvetheorem is named after the mathematician Camille Jordan, who found its first proof[9]. In 1905, O. Veblen declared that this theorem is "justly regarded as a most important step in the direction of a perfectly rigorous mathematics". According to Courant and Robbins, "The proof given by Jordan was neither short nor simple, and the surprise was even greater when it turned out that Jordan's proof was invalid and that considerable effort was necessary to fill the gaps in his reasoning". According to Hales (2007), Jordan's proof is essentially correct... Jordan's proof does not present the details in a satisfactory way. But the idea is right, and with some polishing the proof would be impeccable. In 1978, Dostal and Tindell in their essay "The Jorden Curve Theorem Revisited", wrote "however, notwithstanding substantial simplifications achieved in the elementary proof of JCM by these and other authors, the theorem has remained and will probably always remain, difficult to establish by purely elementary means". This proof was formalized, in 2005, using HOL Light.

(3) The *Odd Order Theorem* precisely says that groups that have an odd number of elements are solvable. Finite groups can be factored somewhat like integers, though for a more complicated multiplication. The basic group factors, called simple groups, also come in many more shapes than the basic integer factors, the prime numbers. Solvable groups, however, can be factored down to primes, like integers. They're called this because they correspond to solvable polynomial equations. Feit and Thompson classically demonstrated this proof in 1963. Recently, it was formalized, in 2012, using *Coq* System.

Discussion

The history of mathematical reasoning began with the attempt to mechanize the thought processes. Aristotle codified syllogisms and Euclid codified geometry. Frege and Peano worked on combining formal reasoning with the study of sets and numbers. David Hilbert worked on stricter formalizations of geometry than Euclid's. All of these efforts were directed towards clarifying what one means by "proof".

In 1940, giant electronic brain catalysed the convergence of three previously disparate areas: theory of axiomatic reasoning, the study of mechanical computation and Psychology of intelligence. Since then, there has been a restless progress of in computer science and artificial agency. What does this has to do with Plato's theory of arithmetic explained previously in this essay? Following his that perspective, a proof is a statement about the reality that satisfies the condition of *Existence*, *Abstractness* and *Independence.* Mathematical objects exist objectively; this means they (1) are not subject to spatiotemporal causality and are intentionality independent; and (2) that to be referred they demand a definition of truth, which is the one and only correct description of reality.

Modern theorem provers seem to satisfy all of the above conditions in the accomplishment of a computer assisted proof. The great enterprise of Artificial Intelligence is to find out what sort of rules could possibly capture intelligent reasoning. If we assume a

[9] For decades, mathematicians generally thought that first rigorous proof was carried out by Oswald Veblen. However, this notion has been challenged by Thomas C. Hales and others.

platonistic perspective, the machine seems to be much able to perform an arithmetic proof. The following table aims to be self-explanatory on that matter:

Formal System	Classical Mathematical
Interactive Theorem Proving	The Human Mind
Symbolic language	Natural language
Computer assisted proof	Classical proof
Inanimate	Animate
Inflexible	Flexible
Mind Independent	Mind depended
Strings of symbols	Statements
Produced by typographical rules	Theorems are proven
Not subject to Spatiotemporal causality	Subject on spatiotemporal causality
Objectivity	Subjectivity
No intuition	Intuition
So trivial, beyond reproach	"Feels right" — not always provable
Astronomical size	Complexity

According to what we asserted about this issue, we may accurately acknowledge that the inferential result or consequence of a computer assisted proof, satisfies Plato's conditions to intelligibles, since a computer assisted proof's satisfies the preconditions:

(i.) $\exists x Mx$ (Existence).
(ii.) Non-spatiotemporal and (therefore) causally inefficacious (Abstractedness).
(iii.) Independent of intelligent agents and their language, thought, and practices (Independence).

Conclusions

For the reasons stated, we may recognize that the conditions for an agent to perform a calculus of reason are pretty much achieved. In fact, a programed inanimate, intelligent agent with symbolic language is able to demonstrate a proof; a computer assisted proof witch uses strings of symbols produced by typographical rules. From a symbolic language, an inanimate, inflexible, and mind-independent agent in the end accomplishes a proof that is nor subject to spatio-temporal causality. It is an objective, non-intuitive proof that is so trivial, it is beyond reproach. This computer assisted proof is, accordingly, a rebellion "against the introduction of visible or tangibles objects into the argument".

The result of such computer assisted proof is an existent, abstract and independent self-representative proof. A mathematical object whose existence is independent of the human mind: the objective mathematical entity, Plato very much aspired for.

References

Balaguer, M. (1998). Platonism and Anti-Platonism in Mathematics. Oxford University Press

Bealer, G. (1993). 'Universals'. *Jornal of Philosophy*, 60 (1):5-32.

Beaney, M. (1997). *The Frege Reader*. Oxford: Blackwell

Benacerraf, Paul, 1973, 'Mathematical Truth', *Journal of Philosophy*, 70(19): 661–679.

Bernays, Paul, 1935, 'On Platonism in Mathematics', Reprinted in Benacerraf and Putnam (1983).

Boyer, C. and Merzbach, U., (1989). *History of Mathematics*. Second edition, John Wiley & Sons, Inc. New York.

Burgess, J. (1983). 'Common Sense and "relevance"'. *Notre Dame Journal of Formal Logic*, Vol. 24, N. 1.

Burgess, J. P. and Rosen, G., (1997), *A Subject with No Object*, Oxford: Oxford University Press.

Burgess, J. P., Rosen, G. (2005). 'A Subject with no Object'. *Canadian Journal of Philosophy*, vol. 30, no. 1.

Dostal, M., Tindell, R. (1978). *The Jordan Curve Theorem Revisited*. Jber, d. Dt. Math-verein. 80 111-128.

Dummett, Michael, (1978), *Truth and Other Enigmas*, Cambridge, MA: Harvard University Press.

Frege, G. (1974). *The Foundations of Arithmetic*, J. L. Austin (trans.), Oxford: Basil Blackwell.

Frege, G. (1956). 'The thought: a Logical Inquiry'. *Mind*, vol. LXV. No. 259.

Frege, G. (1891). 'Funktion und Begriff', Vortrag, gehalten in der Sitzung vom 9. Januar 1891 der Jenaischen Gesellschaft für Medizin und Naturwissenschaft, Jena: Hermann Pohle. Translated as 'Function and Concept' by P. Geach in *Translations from the Philosophical Writings of Gottlob Frege*, P. Geach and M. Black (eds. and trans.), Oxford: Blackwell, third edition, 1980.

Frege, G. (1892). 'Über Sinn und Bedeutung', in *Zeitschrift für Philosophie und philosophische Kritik*, 100: 25–50. Translated as 'On Sense and Reference' by M. Black in *Translations from the Philosophical Writings of Gottlob Frege*, P. Geach and M. Black (eds. and trans.), Oxford: Blackwell, third edition, 1980.

Geuvers, J.H. (2009). Proof assistants : history, ideas and future. *Sadhana : Academy Proceedings in Engineering Sciences (Indian Academy of Sciences), 34*(1), 3-25. *in Web of Science Cited 4 times*

Gödel, K., 1995, 'Some basic theorems on the foundations of mathematics and their implications', in *Collected Words*, S. Feferman et al, ed., Oxford: Oxford University Press, vol. III, 304–323.

Gödel, K. (1986). Collected Works, Vol. I, Oxford: Oxford University Press.

Hales, Thomas C. (2007a), 'The Jordan curve theorem, formally and informally', *The American Mathematical Monthly* 114 (10): 882–894.

Hales, T. (2008). 'Formal Proof'. *Notices of the AMS*, 55, 11.

Hales, T. (2007), "Jordan's proof of the Jordan Curve theorem", *Studies in Logic, Grammar and Rhetoric* **10** (23).

Harrison, J. (2008). 'Formal Proof – Theory and Practice'. *Notices of the AMS*, 55, 11.

Heath, T.(1965). *A History of Greek Mathematics*. Vol. 1. 1965. Oxford university press. Oxford.

Heawood, P. J. (1890). "Map colour theorem". *Quarterly Journal of Mathematics* **24**: 332–338.

Hofstadter, Douglas R. (1979), *Gödel, Escher, Bach: An Eternal Golden Braid*, Basic Books.

Horsten, L. (2015). "Philosophy of Mathematics", *The Stanford Encyclopedia of Philosophy* (Spring 2015 Edition), Edward N. Zalta (ed.), forthcoming URL = <http://plato.stanford.edu/archives/spr2015/entries/philosophy-mathematics/>.

Maddy, P., (1990), *Realism in Mathematics*, Oxford: Clarendon.

Nagel, E., Newman, J. (2001). *Gödel's Proof.* New Youk University.

Maziarz, E. A. and Greenwood, T.(1968). *Greek Mathematical Philosophy*. Frederick Ungar Publishing Co., Inc. New York.

Parsons, C. (1983). *Mathematics in Philosophy*, Ithaca, New York, Cornel University Press.

Plato. *The Republic*. Translated by B. Jowett. 2000. Dover Publications, Inc. New York.

Penrose, R. (2007). *The Road to Reality*, USA: Vintage Books.

Shapiro, S. (1997). *Philosophy of Mathematics: Structure and Ontology*. Oxford University Press.

Veblen, O. (1905). *Theory on Plane Curves in Non-Metrical Analysis Situs*, Trans. AMS, 6, No. 1, 83–98.

Wiedijk, F. (2006). *The Seventeen Provers of the World*, volume 3600 of *lecture notes in Computer Science*, Springer-Verlag.

Acknowledgements

The Portuguese Foundation for Science and Technologies supported this research. I would like particularly to thank professor Reinhard Kahle for his exceptional insights, and the editor, professor Bharath Sriraman.

Meet the Authors (TME2016,vol.13,no.3)

Christopher Lyons is an Assistant Professor of Mathematics at California State University, Fullerton. His primary research interests lie in number theory and algebraic geometry, especially in relation to algebraic surfaces. As both a teacher and researcher, some of his most rewarding undertakings have involved taking a painfully abstract concept from these fields, and finding concrete examples and questions to make the idea seem natural to an outsider. In other words, he believes it's important (and not at all rude!) in mathematics to ask and try to answer the question, "Why would anybody care about this?" In his nonmathematical life, he spends most of time with his wonderful wife and two children. His remaining thoughts and energies often center on reading and coffee.

Audrey Nasar is an Assistant Professor of Mathematics at Borough of Manhattan Community College in New York City. Her research interests lie in discrete mathematics education, specifically in the study of algorithms and their efficiency. Given the impact of computers and computing on almost every aspect of society, the ability to develop, analyze, and implement algorithms has become increasingly important. Her main goal in teaching mathematics is to bring the excitement of problem solving and the beauty of proof into the classroom. Regardless of one's background in mathematics, she believes it is possible to see what has dazzled mathematicians for centuries. Audrey is also fascinated by the intersection of mathematics and art. Her hobbies include printmaking and illustration.

Natanael Karjanto is an Assistant Professor of Mathematics at Sungkyunkwan University, the oldest university in Korea, dating back its history to 1398 during the Joseon Dynasty era. His research interests include Partial Differential Equations, nonlinear wave phenomena and didactics of mathematics, particularly at the tertiary level. He received education in Indonesia and in the Netherlands and has broad international teaching experience, having taught in Malaysia, Kazakhstan and South Korea.

Binur Yermukanova is a senior student majoring in Economics at Nazarbayev University, Kazakhstan. Minoring in mathematics, she has an interest in the sphere of Partial Differential Equations. Performing excellent academic achievement in the Bachelor's degree, she is interested in continuing her education at a graduate level, particularly in the field of Microeconomic Theory.

Milos Savic is an Assistant Professor of Mathematics at the University of Oklahoma. His research focuses on understanding and fostering mathematical creativity in the undergraduate classroom, specifically pertaining to proving. Creativity, he believes, is at the heart of students' difficulties with proving, yet he believes it is one of the most important characteristics of being a mathematician. The creativity collaboration group that he participates in spans the continental United States, and he would be remiss if he did not acknowledge the great work done by his collaborators. He gladly uses all of his non-research time to foster creativity in the house with his wife and two daughters.

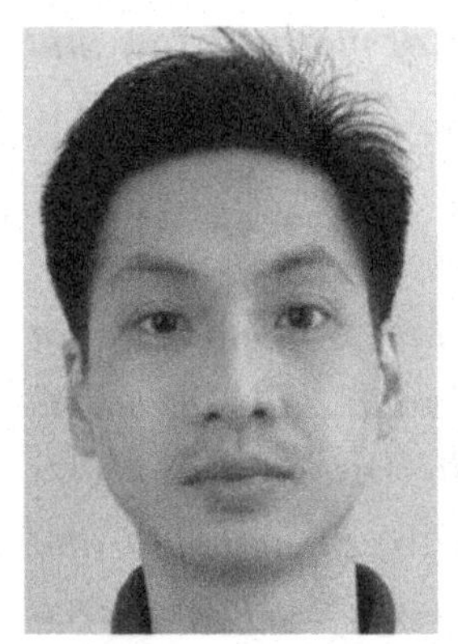

Hartono Tjoe is an assistant professor of mathematics education at the Pennsylvania State University, where he teaches courses in mathematics and mathematics education. He received his PhD in mathematics education from Columbia University. His research focuses on the role of cognitive and mathematical structures in the process of problem solving.

Ines Hipolito is a Philosophy graduated student from Lisbon (BA. MA. PhD ABD). Her main research topics fall under the scope of Philosophy of Mathematics and Logics, Philosophy of Mind, Phenomenology and Cognitive Science, and include subjective experience; perception; 4E cognition; consciousness. Ines is a member of the FCT funded research project Hilbert 24th problem from Nova University of Lisbon, in which her main research interest is the intersection between Hilbert and Husserl's work as a reciprocal and simultaneous internalization of mathematics into philosophy and of philosophy into mathematics. Ines is also a member of the research group Lisbon Mind and Cognition from the Institute of Philosophy of Nova, and a member of the Doctoral College Mind-Brain from Lisbon University.

Printed in the United States
By Bookmasters